Pharmaceutical Toxicology

Safety sciences of drugs

Edited by

Gerard J Mulder

Leiden Amsterdam Center for Drug Resarch (LACDR)
Leiden University
Leiden
The Netherlands

Lennart Dencker

Professor and Chairman of Toxicology and
Dean Faculty of Pharmacy
Uppsala University
Uppsala
Sweden

European University Consortium for Pharmaceutical Research

UPPSALA • LEIDEN • LONDON
AMSTERDAM • PARIS • COPENHAGEN

(PhP)

London • Chicago   **Pharmaceutical Press**

Published by the Pharmaceutical Press

An imprint of RPS Publishing

1 Lambeth High Street, London SE1 7JN, UK

100 South Atkinson Road, Suite 200, Grayslake, IL 60030-7820, USA

© Pharmaceutical Press 2006

(**PP**) is a trade mark of RPS Publishing

RPS Publishing is the publishing organisation of the Royal Pharmaceutical Society of Great Britain

First published 2006
Reprinted 2007, 2008

Typeset by Type Study, Scarborough, North Yorkshire
Printed in Great Britain by TJ International, Padstow, Cornwall

ISBN 978 0 85369 593 6

Pharmaceutical Toxicology

Contents

ULLA postgraduate pharmacy series

The ULLA series is a new and innovative series of introductory text-books for postgraduate students in the pharmaceutical sciences.

This new series is produced by the ULLA Consortium (European University Consortium for Advanced Pharmaceutical Education and Research). The Consortium is a European academic collaboration in research and teaching of the pharmaceutical sciences that is constantly growing and expanding. The Consortium was founded in 1990 and consists of pharmacy departments from leading universities throughout Europe including:

- Faculty of Pharmacy, Uppsala University, Sweden
- School of Pharmacy, University of London, UK
- Leiden/Amsterdam Center for Drug Research, University of Leiden, The Netherlands
- Vrije Universiteit Amsterdam, The Netherlands
- Danish University of Pharmaceutical Sciences, Copenhagen, Denmark
- Faculty of Pharmacy, Universities of Paris Sud, France

The editorial board for the ULLA series consists of several academics from these European Institutions who are all experts in their individual field of pharmaceutical science.

With approximately three books published per year in the pharmaceutical sciences, titles will include:

Pharmaceutical Toxicology
Paediatric Drug Handling
Molecular Biopharmaceutics

The titles in this new groundbreaking series are primarily aimed at PhD students and will also have global appeal to postgraduate students undertaking masters or diploma courses, undergraduates for specific courses, and practising pharmaceutical scientists.

Further information on the Consortium can be found at www.u-l-l-a.org

Preface

This book covers the basics of drug safety sciences and is aimed at students in pharmacy and pharmaceutical sciences at the BSc/MSc level. It focuses more on medicinal drugs than other toxicology textbooks: the most important safety issues of drugs are covered, including registration requirements of new drugs and pharmacovigilance.

The book is not meant to cover every aspect of drug toxicity: we have chosen a limited number of areas that address the major issues and target organs for drug-induced toxicity. Furthermore, we have allowed the chapters to have a somewhat variable approach, some being short general overviews, others including some research aspects. This reflects toxicology and safety sciences of drugs in a broad sense: on one hand it covers routine toxicity screening, on the other it requires in-depth mechanistic investigations in order to understand a certain unwanted drug effect. For instance, drug-induced tumours in a certain organ in rats may need to be investigated mechanistically in order to determine their relevance for human patients and to allow human use of the drug.

After a general introduction to basic toxicological principles, the organ toxicity of drugs is illustrated by chapters on toxicity in the liver, the kidney, the respiratory system and the immune system. Specialised, but highly relevant, issues for safety/efficacy assessment of new drugs or drug indications are teratology and genotoxicity/carcinogenicity. Mechanistic aspects as well as the methodology of safety assessment and pharmacovigilance are dealt with.

For evaluation of the responsible and safe use of a drug, the efficacy/safety balance for each indication has to be judged, and everyone involved in pharmaceutical sciences should be aware of both sides of the balance. This book, therefore, is a good companion to pharmacology textbooks. It combines a broad treatment of the safety issues relevant for assessing the efficacy/safety balance of a new drug with information on the toxicological knowledge base and methodology.

Gerard J Mulder and Lennart Dencker
April 2006

About the editors

Gerard J Mulder is Professor of Toxicology at Leiden University; he was director of research of the Leiden/Amsterdam Center for Drug Research (LACDR), a partner of the ULLA consortium. He obtained his PhD at Groningen University in 1973 and did a post doc with Dr James R Gillette at the National Institutes of Health, Bethesda, MD, USA in 1975/1976. He has published extensively on drug metabolism *in vivo* and *in vitro*, in particular glucuronidation, sulfation and glutathione conjugation: he is author of over 200 publications and has edited several books in this area. Mechanisms of xenobiotic toxicity in liver and kidney, as well as chemical carcinogenesis, are the major subjects of his research interest.

Gerard Mulder is a member of the Dutch Medicines Evaluation Board as well as chairman of the Dutch Expert Committee on Occupational Safety. As such, he is heavily involved in risk assessment of both medicinal drugs and occupational chemicals.

Lennart Dencker is Professor and Chairman of Toxicology and Dean of the Faculty of Pharmacy at Uppsala University, Uppsala, Sweden 1999–2005. He has a DVM from the Royal School of Veterinary College in Stockholm, Sweden (1970) and a PhD in Toxicology from Uppsala University (1976). After a postdoctoral period at National Institutes of Health in Bethesda, MD, USA, he held various teaching and research positions at Uppsala University until he became full Professor and Chairman of the Department of Toxicology in 1986.

Lennart Dencker serves or has served on a number of Swedish committees, such as the National Drug Administration (MPA), National Chemical Inspectorate, National Food Administration, and the Grant Review Committee on Toxicology at the Swedish Environmental Protection Agency, as well as international ones such as the European Science Foundation, Strasbourg (Toxicology Steering Group) and the Norwegian Research Council. He has been Chairman of EUROTOX'93 in Uppsala (1993) and Chairman of the Swedish Society of Toxicology. He has published over 100 papers in international refereed journals.

Contributors

Eva Brittebo, MSciPharm, PhD
Professor in Pharmaceutical Toxicology Department of Pharmaceutical Biosciences; Division of Toxicology, Uppsala University, Uppsala, Sweden

Bengt R Danielsson, MD, MSciPharm, PhD
Professor in Pharmacology and Toxicology at the Medical Products Agency, Uppsala, Sweden; and Professor of Pharmaceutical Toxicology, Department of Pharmaceutical Biosciences; Division of Toxicology, Uppsala University, Uppsala, Sweden

Lennart Dencker, PhD
Professor and Chairman of Toxicology, Dean, Faculty of Pharmacy at Uppsala University, Uppsala, Sweden

Björn Hellman, MSciPharm, PhD
Senior Lecturer and Director of Studies in Toxicology, Associate Professor in Experimental Occupational and Environmental Medicine, Department of Pharmaceutical Biosciences; Division of Toxicology, Uppsala University, Uppsala, Sweden

Ronald Meyboom, MD, PhD
Medical Adviser, The WHO Uppsala Monitoring Centre, Uppsala, Sweden; and Senior Researcher, Department of Pharmacoepidemiology and Pharmacotherapy, Faculty of Pharmaceutical Sciences, University of Utrecht, Utrecht, The Netherlands

Gerard J Mulder, PhD
Emeritus Professor of Toxicology, Leiden Amsterdam Center for Drug Research (LACDR), Leiden University, Leiden, The Netherlands

J Fred Nagelkerke, PhD
Associate Professor at the Division of Toxicology of the Leiden Amsterdam Center for Drug Research (LACDR), Leiden University, Leiden, The Netherlands

Sten Olsson, MSciPharm
Head, External Affairs, The WHO Uppsala Monitoring Centre, Uppsala, Sweden

Hans Persson, MD, MDhc
Consulting Physician in Clinical Toxicology, Swedish Poisons Centre, Stockholm, Sweden

Camilla Svensson, MSciPharm, PhD
Faculty of Pharmacy, Department of Pharmaceutical Biosciences; Division of Toxicology, Uppsala University, Uppsala, Sweden

Bob van de Water, PhD
Professor of Safety Sciences/Toxicology of the Leiden Amsterdam Center for Drug Research (LACDR), Leiden University, Leiden, The Netherlands

Jan Willem van der Laan, PhD
Institute for Public Health and the Environment, Bilthoven, The Netherlands; and the Dutch Medicines Evaluation Board, The Hague, The Netherlands

1

General toxicology

Björn Hellman

Toxicology is an essential part of the development process of new drugs in the pharmaceutical industry because ultimately the balance between safety and efficacy has to be established. The resulting information on drug-induced toxicity, including different types of side-effects and inter-actions, is of great concern for consumers of drugs as well as for pharma-cists, healthcare practitioners, agencies regulating medicinal products and others that have responsibilities in different aspects related to the safe use of drugs. A lot of toxicological information can now be found easily on the Internet; unfortunately, this information is not always accurate. Moreover, knowledge about fundamental principles in toxicology is needed in order to evaluate even the most reliable information.

The fact that most toxicological data derive from studies on experimental animals reinforces the importance of knowledge of the premises for toxicity testing, as well as the way in which toxicological data are used in safety assessment. When evaluating the 'toxicological profile' of a chemical, information is gathered about its rate and pattern of absorption, distribution, metabolism and excretion ('ADME'), as well as its immediate and delayed adverse health effects, target organs of toxicity, clinical manifestations of intoxication, mechanism(s) of action and dose–response relationships.

This chapter focuses on the basic principles in toxicology neces-sary for understanding how toxicity data are used in safety assessment of drugs for human use. Since quantitative aspects of dose–response relationships and pharmacokinetics of toxicants ('toxicokinetics') are on the whole very similar to those in pharmacology, those aspects are treated only briefly, outlining relevant differences where appropriate.

The concept of toxicity is not easily defined

Toxicology (the science of 'poisons') deals with chemically induced adverse effects on living organisms. These chemicals ('toxicants') include both synthetic agents ('xenobiotics' or 'foreign compounds') and naturally occurring substances such as the poisons produced by bacteria, animals and plants (often referred to as 'toxins' or 'venoms'). Toxicology is a multidisciplinary science applying methods and traditions from several other disciplines (biochemistry, cell biology, pathology, pharmacology, physiology and analytical chemistry). The mainstream of toxicology focuses on describing and evaluating toxicity from the human health perspective, and safety assessment of drugs aims to predict human health hazard and risks.

Toxicity is often defined as the intrinsic ability of an agent to harm living organisms. This definition is not unequivocal because it will ultimately depend on how 'harm' is defined. Toxicity can also be defined as an adverse health effect associated with a change, reduction or loss of a vital function. This includes an impaired capacity to compensate for additional stress induced by other (e.g. environmental) factors. For example, many survivors of the Bhopal disaster in 1984 (in which severe lung toxicity occurred owing to the accidental release of methyl isocyanate) had a reduced lung function that made them more sensitive to otherwise mild infections. Although the accident occurred over 20 years ago, people may still die because of that tragic incident.

Clearly, drugs can induce a broad spectrum of undesired health effects, some of which are clearly deleterious, others that are not. In safety evaluation of drugs, toxicologists generally focus on direct adverse effects upon an exposure to therapeutic doses. But harmful effects, for example malformations or even death, may also be the result of an indirect effect such as drug-induced deficiency of an essential element (such as vitamin A or selenium). In addition, toxic agents (including drugs) may interact, which can result in both increased and decreased responses.

Changes in morphology, physiology, development, growth and lifespan leading to impairment of functional capacities are typical examples of 'toxic', 'deleterious', 'detrimental', 'harmful', 'injurious', 'damaging', 'unwanted', 'adverse' or 'side' effects. But should a subtle change in the blood pressure or a small change in a subset of lymphocytes be regarded as adverse effects? They could equally well be considered as 'just' biological indicators of exposure if they are (rapidly) reversible.

In conclusion, whereas severe adverse health effects are easily defined, there is also a 'grey zone' of effects of doubtful significance in terms of human health.

Each drug has a unique toxicological profile

Drugs (which are often classified in terms of their therapeutic use) include many different types of agents, producing different types of adverse effects by various mechanisms of action. Therapeutic agents belonging to a given class of compounds often have some adverse effects in common ('class effects'). For instance, nonsteroidal anti-inflammatory drugs have gastrointestinal side-effects in common and may also be nephrotoxic, but as a rule each individual compound should be expected to have its own unique 'toxicological profile'. Since chemical and physical properties of the compound (water solubility, hydrogen bonding, reactivity, size, degree of ionisation, etc.) play an important role in the expression of this profile, knowledge of these properties is a prerequisite when testing and evaluating the toxicity of a chemical.

The biological effects of a drug are usually a function of the chemical structure of the parent compound or its metabolites, and the chemical and physical properties will to a large extent determine whether an agent will induce either local or systemic adverse effects. Whereas most drugs express their effects after they have been absorbed and distributed in the body (systemic effects), some chemicals (e.g. strong acids and bases, or highly reactive compounds such as epoxides) act primarily at the first site of contact. Typical examples are the severe burns to eyes and the skin following splashing of a strong alkaline agent in the face, the ulcers in the epithelia of the digestive system following ingestion of a corrosive agent, and the inflammatory reactions in the respiratory tract following the inhalation of an irritant agent.

In conclusion, each compound should be expected to have its own characteristic toxicological profile (including class effects); the route of exposure and physicochemical properties of the compound to a large extent determine the site of toxicity.

Is it only the 'dose' that makes the poison?

One of the most fundamental concepts in toxicology is that it is the dose that makes the poison. This means that most chemicals will become toxic at some dose. Whereas some compounds are lethal if ingested in minute quantities (e.g. botulinum toxin), others will induce their adverse effects only if ingested in relatively large quantities (e.g. saccharin). In most cases, a chemical cannot induce any adverse effects unless it reaches a critical site at a sufficiently high concentration for a sufficiently long period of time. From this it follows that even an extremely toxic substance will be harmless as long as it is kept in a closed container, and that a relatively non-toxic chemical can be rather hazardous if handled carelessly.

Most drugs are taken orally and typical measures for the dose are then mg/kg or µmol/kg body weight (or per cm^2 body surface area for interspecies comparisons). The same measures are used for intravenous injections and any other type of bolus dose. For inhalation experiments using sprays, it is not only the dose but also the concentration that is of importance, and the same applies for drugs administered on the skin. If a small amount of a drug is given at an extremely high concentration, the drug may have a strong effect locally (causing, for example, severe erythema), but if the same amount is administered in a much more diluted form it may not cause any local reactions at all.

Often it is more informative to talk about *dosage* than dose. The dosage can be defined as the amount of toxicant taken by, or given to, the organism over time (for example, in a repeat-dose toxicity study) and the typical measure is then mg/kg body weight per day. An even better measure of the actual exposure is the internal (systemic) dose, because this way of expressing the exposure is more directly related to the potential adverse health effects than is the dose or dosage (at least for toxicants with systemic effects; see below). The internal dose (usually the concentration of the toxicant in the blood) is therefore regularly monitored in toxicity studies of drugs. Choosing the appropriate dosage is very important when designing a toxicity study (especially a long-term study). Critical health effects may be overlooked if the dosage is too low. If the dosage is too high, this may lead to early deaths, which complicates the analysis of the study, especially when the interpretation of the outcome is dependent on a reasonable survival of the animals such as in 2-year carcinogenicity studies in rodents.

Clearly, the concentration of the toxicant at the site of action is related to the dosage. The final 'target dose' (i.e. the amount of toxicant present at the critical site for the necessary period of time) is governed by several factors such as the exposure situation and the fate of the toxicant in the body once it has been absorbed. There can be big differences in susceptibility between individuals (and species) exposed to a particular toxicant. Intra- and interindividual variations in susceptibility depend on several factors such as genetic constitution, age and sex, health condition and nutritional status, previous and ongoing exposures to other toxicants, and climate conditions. All these factors should be considered when using data obtained under one set of conditions to predict what the outcome would become under another set of conditions.

In conclusion, although the concept of 'dose' looks quite simple, it is not easy to define unequivocally. It can relate to the 'external dose' (the amount actually ingested, inhaled or applied on the skin), the 'systemic (or internal) dose' (usually the concentration in blood), the 'tissue (or organ) dose' (the amount or concentration of the toxicant in various tissues after absorption, distribution and metabolism), or the 'target dose' (the amount of the ultimate toxicant actually present at the critical site for a sufficient period of time). Tissue and target doses are often very difficult to measure (especially in humans), so the systemic dose is usually the most precise measure of exposure in relation to the risk for adverse health effects.

Drugs can induce both immediate and delayed toxicity

In the 'good old days', a compound was often considered harmless if it was without immediate adverse health effects when administered in a large single dose. Nowadays it is recognised that some toxicants accumulate in the body and that the 'tissue doses' will eventually become critically high if the exposure to such agents continues for a sufficiently long time, even at rather low doses. It has also been recognised that a short-term low-dose exposure to some types of toxicants (e.g. potent genotoxic agents) may be sufficient to induce delayed adverse effects (malignant tumours).

The terms 'acute' and 'chronic' are used to describe the duration and frequency of exposure in toxicity tests, but these terms can also be used to characterise the nature of the observed adverse health effects. Consequently, although a single dose exposure in most cases is associated with acute effects (i.e. immediately occurring adverse effects manifested within a few minutes up to a couple of days after the exposure), it can also induce delayed adverse effects manifested only after quite some time. One obvious example of a delayed effect following from an acute exposure is the lung cancer following from an acute inhalation of plutonium-239 (in the form of a Pu^{4+} salt). A less obvious, but still striking, example is the delayed neurotoxicity that was observed in humans who were exposed to the organophosphorus ester tri-*ortho*cresylphosphate (TOCP). In the latter case, severe axonopathy was observed both in the central and peripheral nervous systems several days after an accidental acute oral exposure without any signs of cholinergic poisoning, the immediate acute effect typically induced by many other organophosphorus esters. Long-term chronic exposures are usually associated with chronic effects.

Depending on the duration and frequency of exposure, experimental studies on the toxicity of chemicals are usually referred to as either short-term or long-term toxicity studies (chronic studies). The route of administration of a drug in such tests should always include the intended human route of administration. The maximum duration of exposure in an acute study is usually limited to 24 hours. The compound may be administered orally (in most cases as a single dose), by inhalation (e.g. for 6 hours) or cutaneously (usually for 24 hours on a shaven area of the skin). The maximum duration of exposure in a short-term repeated dose study (formerly referred to as a 'subacute' study) is limited to one month. In a subchronic toxicity study, a period up to 10% of the normal lifespan of the animal is used (usually 90 days for rodents).

The duration of exposure in a long-term toxicity study should be at least 12 months, but is usually 18–24 months for mice and rats. In the long-term toxicity studies, the test compound can be given via the diet or administered in the drinking water (continuous exposure), by gavage or capsule (usually one oral dose/day, 5 days/week), on the skin (usually one daily application, 5 days/week), or in the inhaled air (e.g. 8 hours/day, 5 days/week). In some studies, the animals are exposed for several generations (e.g. in two-generation reproduction toxicity studies).

Bioavailability and toxicokinetics: two important factors for systemic toxicity

Bioavailability represents the extent to which the chemical reaches the systemic circulation from the site of administration. Maximum bioavailability (and therefore the most intense and rapidly occurring toxic response) results after an intravenous injection when the bioavailability is by definition 100%. The route of entry along which a compound enters the body is a decisive factor determining the bioavailability. Chemical and physical properties determine its (rate of) absorption and first-pass metabolism. After oral administration, the bioavailability can be close to 100% for completely absorbed agents without first-pass metabolism, but usually it will be less.

Another important factor is the rate at which the toxicant is released from its environmental matrix (from nanoparticles, inhaled particles, slow-release preparations, creams, etc.). Since most toxicants are absorbed by simple diffusion, small, lipid-soluble and non-ionised molecules will in general be more readily absorbed (i.e. have better bioavailability) than bulky, less lipid-soluble, ionised molecules.

Studies on the rates and patterns of absorption, distribution, metabolism, and excretion (ADME) of toxicants are known as toxico-kinetic studies. They are essential for assessing the systemic exposure to a drug and its metabolites, because what is in the blood will reach the tissues. When studying the toxicokinetics of a chemical in experimental animals, the compound can be administered either as it is, or labelled with a radioactive isotope (usually tritium (^{3}H) or carbon-14 (^{14}C)). The concentration of the toxicant (and/or its metabolites) is then usually determined after various intervals in different body fluids, organs and/or excreta, using gas or liquid chromatographic methods and/or mass spectrometry. Toxicokinetic studies should be performed using both high and low doses, single and repeated exposures, different routes of exposures, both sexes, different ages, pregnant and non-pregnant animals, and different species. Knowledge of the 'fate' of a toxicant in the body under different exposure conditions facilitates the selection of appropriate testing conditions when designing the subsequent toxicity studies.

The *kinetic* parameters determined in toxicokinetic studies are used in mathematical models to predict the time course of concentration of the toxicant (and/or its metabolites) in various 'compartments' of the organism. By using 'compartmental' or 'physiologically based' models it is, for example, possible to calculate various absorption and elimina-tion rate constants, hepatic, renal and total body clearances, biological half-lives, apparent volumes of distribution, and steady-state concentra-tions of the toxicant in various organs. This is comparable to the pharmacokinetic approach and methodology.

Knowledge of the internal (systemic) exposure is essential when evaluating and comparing the toxicity of a given compound between different species (including humans). Toxicokinetic studies are therefore crucial when extrapolating animal toxicity data to assess human health risks. They will provide information to assess or predict, for example, possible interactions with various receptors and/or enzyme systems under different exposure conditions for different species. Consequently, estimates of margins of safety (see below) for adverse effects observed in animal studies are more reliable if they are based on toxicokinetic data on systemic exposures of a given drug rather than on the adminis-tered doses, when comparing the systemic toxicity between experimen-tal animals and humans.

In conclusion, the intended route of administration in patients must always be included when establishing the toxicological profile of a particular drug. In general (but not always) the bioavailability for a given dose of the drug decreases in the following order: intravenous injection > inhalation > oral administration > dermal application.

Absorption

There are several barriers a toxicant may have to pass before it is taken up into the blood and can induce its systemic toxicity: the skin, the lungs and the alimentary canal offer biological barriers after dermal, inhalatory or oral administration, respectively. Obviously these barriers are by-passed after an intravenous or intramuscular injection.

Some compounds enter the body by specialised transport systems (e.g. carriers for uptake of nutrients, electrolytes and other essential elements) but most toxicants appear to be absorbed by simple diffusion through the epithelial cell layers in the gut, lung or skin. Small, lipid-soluble and non-ionised molecules are therefore more readily absorbed than bulky, less lipid-soluble, ionised molecules. Very poorly lipid-soluble, highly water-soluble compounds are badly absorbed from the gut, and this may also be true for extremely lipid-soluble compounds because of their poor water solubility, both in the gut lumen and in blood. If the physicochemical properties of a compound are such that it is hardly absorbed from the gut, it will most likely not be able to induce any systemic toxicity. An example of such a drug is orlistat, a lipase inhibitor that acts on gut lipases to achieve weight reduction. Since its therapeutic action should occur only locally inside the gut lumen, it is an advantage that it is not absorbed into the systemic circulation.

Some substances given orally will hardly reach the general circulation because they are metabolised by enzymes in the intestinal mucosa or the liver. If an ingested compound is absorbed in the gastrointestinal tract, it will first pass to the liver through the portal vein, where it may be taken care of by various enzymes (the so-called first-pass effect). If the same substance enters the body via the lungs or through the skin, it will be taken up by the general circulation and may induce systemic toxicity if it is accumulated in sufficiently high concentrations.

There are other, internal, barriers that a toxicant may have to pass before it can induce its toxicity. The most important is probably the 'blood–brain barrier', but there are also other barriers such as the blood–testis barrier. These barriers are formed by sometimes highly specialised cell layers that – unless there are active transport mechanisms available – prevent or impair the penetration of compounds with low lipid solubility.

Carriers in the cellular membranes, especially the P-glycoprotein pump, play a vital role in the maintenance of various barriers in the body that protect sensitive organs from the potential toxicity of different compounds. The blood–brain barrier is a permeability barrier that limits the influx of circulating substances to the brain. It is based on a number of anatomical and physiological characteristics (including pumps, tight junctions between the endothelial cells and a lack of fenestrations) that make this barrier impermeable to many toxicants (except small and lipophilic ones). The essential nutrients and components needed for brain function are usually provided by carrier-mediated uptake. Whether a toxicant can pass the blood–brain barrier will to a large extent decide whether this agent will induce CNS toxicity. The permeability of the blood–brain barrier in the embryo or fetus is not very tight, contributing to a more vulnerable CNS in the fetus/neonate than in the adult.

Formerly it was generally believed that the placenta protects the embryo and fetus from toxicants, but this 'barrier' will in fact only delay the passage of most drugs. Most toxicants will readily pass across the placenta, usually by passive diffusion but in some cases also by active transport; therefore, after some time, the mother and the fetus will have the same internal exposure. The blood–testis (or Sertoli cell) barrier is also believed to offer some protection of male germ cells (during their meiotic and post-meiotic stages), and female germ cells are possibly protected by the so-called zona pellucida surrounding the oocytes. However, the true effectiveness of these barriers is still uncertain.

Distribution

Although some locally induced adverse health effects may indirectly lead to systemic effects (e.g. the kidney damage following severe acid burns on the skin), systemic toxicity cannot be induced unless the toxicant (and/or its toxic metabolites) is present in sufficiently high concentrations in the target organs. Studies of the distribution of a toxicant deal with the process(es) by which an absorbed toxicant (and/or its metabolites) circulates and distributes in the body. Three different types of distributions are of interest: within the body, within an organ, and within a cell.

If a compound is labelled with a radioactive isotope, it is possible to study its distribution using whole-body autoradiography and/or microautoradiography. The concentration of an unlabelled test substance (and/or its metabolites) can also be measured using various analytical methods. If a particular organ accumulates the drug or its metabolites, this may raise concern. Some examples of high and selective accumulation of drugs elegantly demonstrated by whole-body autoradiography are the accumulation of chlorpromazine and chloroquine in the uvea (pigmented layers behind the retina of the eye), that of thiouracil in the thyroid gland of both the adult animal and the fetus, and that of gentamicin in the proximal tubular kidney cells.

After absorption has taken place and the compound has entered the blood, it is usually distributed rather rapidly throughout the body. The rate and pattern of distribution depend on several factors, including the solubility of the drug in the blood, the regional blood flow, the affinity of the toxicant to various serum proteins and tissue constituents, and carrier-mediated uptake by certain cells (e.g. the proximal tubular cells in the kidney). Whereas some toxicants accumulate in their target organs (e.g. chloroquine in the uvea), others will concentrate in tissues not primarily affected by toxicity. Highly lipid-soluble drugs, for example, accumulate in fat depots, resulting in a large volume of distribution and a long half-life of elimination but no local damage.

Biotransformation

Since the biotransformation of xenobiotics plays a major role in their detoxification as well as their bioactivation to toxic metabolites (i.e. toxification), the rate and pattern of biotransformation is one of the most critical factors determining whether a given chemical will be able to induce toxicity. A number of factors influence the biotransformation of a toxicant, such as genetic constitution, age, sex, species, strain, nutritional status, underlying diseases and concomitant exposures to other xenobiotics with enzyme-inducing and/or enzyme-inhibiting activities.

During their evolution, mammals have developed rather specialised systems to deal with the plethora of foreign substances that enter the body every day. Biotransformation converts the xenobiotics to more water-soluble products so that they can be more readily eliminated from the body via the urine and/or faeces. Biotransformation is often assumed to be detoxification, leading to an increased elimination of less-toxic metabolites. However, sometimes this process can also lead to bioactivation. For a more detailed survey of the role of drug metabolism, see Chapter 2.

Excretion

Elimination of a toxicant from the body is usually studied by measuring the amount of the toxicant and/or its metabolites in the excreta (typically urine, faeces and/or expired air). These measurements are usually performed until approximately 95% of the administered dose has been recovered. The kidney is the most important organ for the elimination of xenobiotics and their metabolites. The processes of elimination via the kidneys are rather complex, but there are at least three different pathways that are of interest: glomerular filtration, tubular excretion by passive diffusion, and active tubular secretion (mainly for organic acids and bases and some protein-bound toxicants). Compounds that are small enough to be filtered with the plasma in the glomeruli can be reabsorbed in the tubuli if they are sufficiently lipid-soluble (see Chapter 8).

Non-absorbed ingested materials as well as compounds excreted in bile from the liver or by intestinal cells in the gut wall are excreted in the faeces. Biliary excretion is an important route of elimination for some toxicants. When drugs are metabolised in the liver, the metabolites can either enter the bloodstream (and be excreted by the kidney) or be excreted into the bile (usually by carrier-mediated transport). Compounds with a fairly high molecular weight (above 300–500 in rats, guinea pigs and rabbits, and above 500–700 in humans) are primarily excreted in bile; those with lower molecular weights are excreted in the urine. Compounds excreted in the bile can also be reabsorbed from the small intestine (enterohepatic recirculation). Pulmonary excretion can be an important route of elimination for volatile compounds. The rate of elimination via the respiratory system depends on several factors, the most important ones being the volatility of the compound, its solubility in the blood, the blood flow to the lungs, and the rate of respiration.

In order to recover 95% of the administered dose, it may sometimes also be necessary to measure the amounts in breast milk, sweat, saliva, tears and hair. Excretion in breast milk is obviously important for mothers breast-feeding their baby because the baby may be exposed to a drug taken by the mother and/or to its metabolites. In the case of persistent environmental chemicals, this is an important route of unwanted exposure.

A toxicant will accumulate in the body if the rate of absorption exceeds the rates of biotransformation and/or elimination. The biological half-life ($t_{1/2}$) of a compound can be defined as the time needed to reduce the absorbed amount in the body by 50%. More commonly it represents the elimination half-life of the concentration of toxicant in plasma (or in a specific organ). As in pharmacokinetics, a time equal to several half-lives may be required to characterise the complete elimination of a compound. Obviously, long half-lives may lead to prolonged toxic effects. The biological half-life varies considerably between various toxicants, from hours (e.g. for phenol), to years (e.g. for some dioxins) or even decades (e.g. for cadmium).

Dose–response and thresholds: fundamental concepts in toxicology

As in pharmacology, there is a quantitative relationship between the 'dose' (the magnitude of exposure) and the 'toxic response' (the magnitude of the induced adverse effect). It is often useful to make a distinction between a 'dose–effect' relationship, i.e. the graded response after exposure to varying doses of a toxicant, and a 'dose–response' relationship, i.e. the incidence at which a 'quantal' response occurs in a population exposed to varying doses of the toxicant (Figures 1.1 and 1.2). Death and clinically manifest tumours are obviously quantal responses ('all-or none' effects), but all types of adverse effects can be converted into quantal responses if a cut-off procedure is used when distinguishing between a defined adverse effect level and a no-effect level (Figure 1.2). Consequently, whereas the dose–effect relationship expresses the extent of an effect in an individual or a group (e.g. the extent of liver toxicity expressed as the activity of transaminases in the blood, or the bilirubin level), the dose–response relationship expresses the incidence of a specific effect at a population level.

When an adverse effect is reproducibly and dose-dependently observed *in vitro* or *in vivo* (in animals or humans), there is a *causal* relationship with the (internal) exposure. This relationship can be quantitatively characterised in toxicokinetic and toxicodynamic studies. This implies that the adverse effect is induced by the toxicant and not by some unrelated factor. Unless the toxicant acts by affecting the microenvironment in the cells or tissues (see below), it is also usually implied that there is some kind of 'critical receptor' with which the toxicant interacts in order to induce its toxicity, that the effect or response magnitude is related to the concentration of the toxicant at this 'receptor site', and that this receptor concentration is related to the administered dose.

One example of toxicity following from a critical change in the microenvironment of cells or tissues, rather than a direct interaction between the toxicant and a specific endogenous target molecule, relates to the adverse effects induced by ethylene glycol through its metabolite

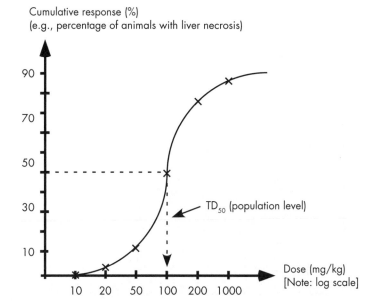

Figure 1.1 Cumulative dose–response relationship. When animals are treated with increasing doses of a drug that causes, for instance, liver necrosis, the percentage of the animals that show an effect at each dose can be determined and plotted against the log of the dose. Experimental points are indicated in this sigmoid plot. The TD_{50}, the dose at which 50% of the treated animals show the predefined effect, is indicated. The predefined effect in this case may be that more than 10% of the cells on histological slides of liver samples taken 12 hours after treatment are necrotic. The threshold is around 10 mg/kg.

oxalic acid. This compound can induce hypocalcaemia due to calcium chelation; it forms water-insoluble calcium oxalate crystals, which are deposited in the kidney tubuli and in small blood vessels of the brain (i.e. in organs known to be damaged by ethylene glycol). Other examples include compounds that alter the hydrogen ion concentration in the aqueous biophase of the cells (e.g. acids shifting the pH) and agents that, in a non-specific way, alter the lipid phase of cellular membranes (e.g. detergents and organic solvents).

Dose–response curves can be constructed in many different ways

Figures 1.1–1.7 show different types of relationships between 'dose' and 'response' or 'effect'. The magnitude of exposure (the independent

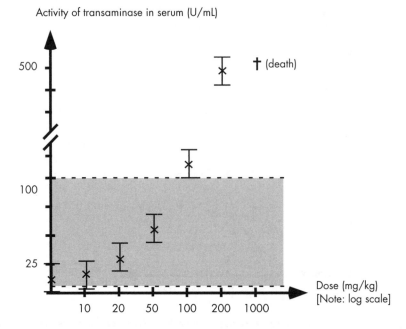

Figure 1.2 Dose–effect relationship. Groups of animals were treated with increasing doses of a hepatotoxic drug and 24 hours after treatment the activity of transaminases was measured in serum. The mean activity per group is given (with standard deviations). The grey zone indicates the range of activities in historical controls (normal range) and activities above that are considered positive (in this case they indicate liver toxicity). To convert the graded response to a quantal response, animals with an activity above the grey level can be counted as positive in the choice: yes/no toxicity.

variable) is usually plotted semi-logarithmically on the x-axis, and the magnitude of response or effect (the dependent variable) on the y-axis. A major reason for using a log scale on the x-axis is that this produces a symmetrical curve and allows a broader range of doses on the graph. A dose–response curve (with a 'quantal response') shows which portion of the population will respond to a given dose of a toxicant. Many different units can be used on the y-axis, including cumulative percentages (as in Figure 1.1), frequencies (as in Figure 1.3), or so-called probit units (as in Figure 1.4). The units used on the y-axis in a dose–effect curve where the response (effect) varies continuously with the dose are usually arithmetic and can be actual (e.g. U/ml as in Figure 1.2) or derived (e.g. percentage of maximum response).

To establish a dose–response curve, the minimum dose required to

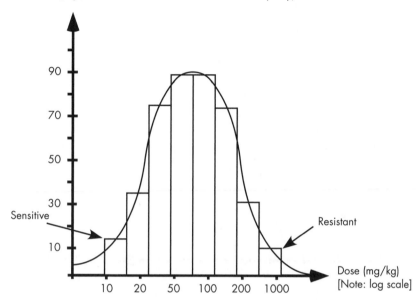

Figure 1.3 Dose–response relationship: noncumulative. In individual animals the occurrence of a toxic effect (e.g. neuronal axonopathy) was 'titrated', so that for each individual animal the dose could be established at which a predefined pain effect occurred. The animals were subsequently grouped together in dose groups (the bars in the figure). The curve shows the sensitive individuals at the left and the resistant individuals at the right-hand side. When the bars are added up (accumulated), this results in the curve of Figure 1.1.The log normal distribution approximates a Gaussian distribution (i.e. it is normally distributed). Then a frequency response can be expressed in multiples of standard deviation (normal equivalent deviates). In such a case, the dose interval at which 50% of the population will respond ±2 standard deviations will, for example, include 95% of the exposed population. The normal equivalent deviates are often transformed into probit (probability) units by adding a constant value of 5 to the standardised normal distributions (to avoid negative numbers) (see also Figure 1.4).

Gaussian distribution (normally distributed data)

Per cent responding	Normal equivalent deviate (SD)	Probit unit
99.9	+3	8
97.7	+2	7
84.0	+1	6
50.0	0	5
16.0	−1	4
2.3	−2	3
0.1	−3	2

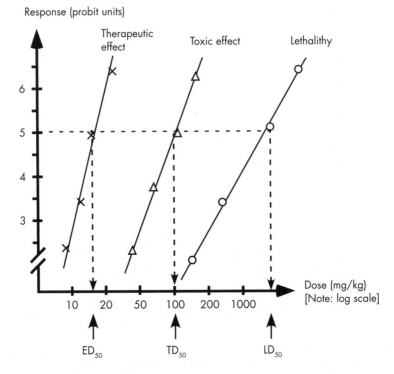

Figure 1.4 Probit plot of various effects. Dose–response data can be plotted using a probit scale to extrapolate data: probability units obtained from standardised normal distributions are plotted against the logarithm of the dose of a substance when quantal (or graded) responses have been measured (in this case a therapeutic effect, a toxic effect and death, respectively). Log-dose data provide linear plots, which make it easy to extrapolate the data. ED_{50} = the dose that produces the desired effect in 50% of the population; TD_{50} = the dose that produces the defined toxicity in 50% of the population, and LD_{50} = the dose that produces death in 50% of the population.

produce a predefined effect (e.g. a liver toxicity defined as 4 times the mean control level of transaminase activity in blood) in each individual of a group should be determined. If the population is very homogeneous (inbred rats without differences in lifestyle), this minimum dose will vary very little between the individual rats; but in a human population genetic and lifestyle differences will result in widely different sensitivities. To obtain this curve, one could titrate each individual of the entire population with increasing doses up to the dose that shows the predefined effect in each individual. The required dose for each individual is noted and the individuals are combined into dose groups to make a plot as in

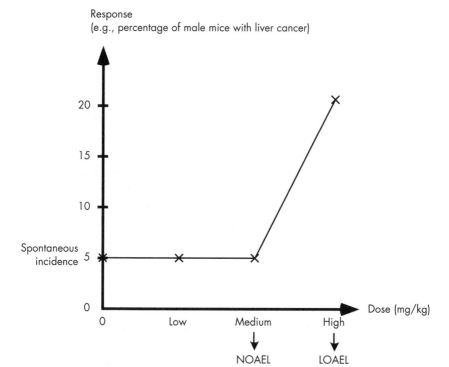

Figure 1.5 Dose—response for a carcinogenic effect. A typical dose—response curve obtained in a repeat dose toxicity study for carcinogenicity employing a control group (usually given vehicle only) and three groups of animals exposed to different doses of the test compound. NOAEL represent the 'no observed adverse effect level', and LOAEL the 'lowest observed adverse effect level'.

Figure 1.3. A more practical way is to randomly divide the test population into groups and give each group only one of a series of increasing doses. The response (percentage of responding animals) in each group is then recorded and plotted against the dose as in Figure 1.1.

The only requirements when establishing a dose—response curve in an experimental study (Figure 1.5) are that the toxicant can be administered accurately, and that the means of expressing the toxicity are precise. As indicated above, dose—effect relationships show the correlation between the extent of an effect (e.g. a decrease in respiratory volume) and the dose, either for an individual or for the group. In the latter case the mean of the individual responses is given, but this can of course only be done when the group is relatively homogenous (i.e. when the individual curves are distributed according to the Gaussian curve).

This is usually the case in experimental studies using inbred animals, but not necessarily in studies employing a group of human subjects. If such a group is very heterogeneous and is composed of two (or more) subgroups with pronounced genetic or lifestyle differences (for example in their metabolism) such means are rather useless.

A very sensitive subgroup may be defined as a 'special group at risk'. The individuals in such a group will, for various reasons, show an excessive toxic response to a certain exposure as compared to the majority of individuals in the general population. For instance, when asthmatic persons are exposed to an airway irritant or histamine at a low dose level that only marginally affects a healthy individual, they can experience a very severe, potentially life-threatening response. Clearly, a mean dose–effect curve for both groups together would not represent either group: each group would show a separate Gaussian distribution of sensitivities (Figure 1.3) peaking at different doses.

Thresholds or not?

Dose–effect relationships are closely associated with a critical issue in toxicology – that of whether or not thresholds exist for various types of adverse effects. Below the threshold (a dose, dosage, concentration, exposure level, etc.) there should be no adverse health effects, but toxicity may occur above this 'minimally effective dose' (threshold). Obviously, the threshold level for a chemical may be different for various individuals and it is also difficult to establish precisely. This is a highly relevant issue when discussing 'safe' exposure levels. The therapeutic effect of a drug always requires a certain minimum dose, which may vary between individual patients. Similarly, the dose at which side-effects will occur can be quite different between different patients. A very sensitive patient may experience unacceptable adverse health effects at a therapeutically required dose of a drug that is safe for the majority of patients, implying that the safe use of this drug is impossible for this particular patient. This mainly applies to side-effects/adverse health effects that operate by a mechanism other than that of the therapeutic effect. Side-effects that are due to the same receptor as the therapeutic effect ('pharmacological side-effects') will be expected to follow the sensitivity for the therapeutic effect.

Most acute and chronic adverse effects are associated with thresholds. The threshold for a particular adverse effect will be determined by characteristics of each individual. In an inbred species (of experimental animals), each individual will show little variation in threshold, but between species or even between different strains of one species there may be pronounced differences. As indicated in Figure 1.5, dose–response curves can be used to determine a 'no observed adverse effect level' (NOAEL) and a 'lowest observed adverse effect level' (LOAEL). The threshold is somewhere in the interval between NOAEL and LOAEL. It should be pointed out that the NOAELs and LOAELs are not absolute effect levels; they are highly dependent on the testing conditions, such as the number of animals in each group, the dose intervals used in the study (which are decided by the study director), the methods for registration of responses, the number of surviving animals, the number of animals and organs subjected to histopathological examinations, and so on.

For some type of adverse health effects (notably neoplasms and genetic diseases induced by mutagens interacting directly with the genetic material, and possibly also sensitisation), it cannot be excluded that there is no true threshold (Figure 1.6). At least theoretically, a single molecule may be sufficient to induce the critical response. 'The black box' of toxicology in Figure 1.6 refers to the true nature of the dose–response curve at very low doses (close to zero) which cannot realistically be studied experimentally (because of the extremely low response).

The 'black box' of toxicology

Quantitative risk assessments concerning exposures to extremely low environmental concentrations of genotoxic carcinogens have to deal with this 'black box', because regulatory actions to reduce such an exposure may be initiated at risk levels as low as 1×10^{-6} to 1×10^{-5}. An extra cancer risk of 1×10^{-6} is equivalent to one extra case of cancer among one million exposed individuals (over a lifetime exposure) or, expressed on the individual level, an extra cancer risk of 0.0001%. This obviously cannot be determined experimentally. This is seldom an issue for the safety evaluation of drugs because most exposures that may lead to adverse health effects occur at therapeutic doses, far above the 'close to the zero dose' region (see below).

The continuing discussion of whether thresholds exist for genotoxic agents involves questions about, for example, the efficiency of detoxification at very low exposure levels and the importance of various DNA-repair pathways (e.g. the balance between the error-free, high-fidelity excision repair pathways and more error-prone pathways). In fact there is a constant and unavoidable exposure to 'endogenous

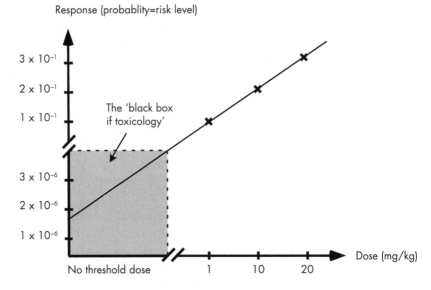

Figure 1.6 Extrapolation from high to extremely low exposure for carcinogens. To calculate the risk for the general population of exposure to a carcinogenic compound, the data obtained at high exposure in experiments in, say, rats (when, for example, 20 of the 50 treated animals develop tumours compared to 3 per 50 in the controls) have to be extrapolated to the much lower exposure at which an incidence of perhaps one additional case of chemically induced cancer per 1 000 000 (risk level: 10^{-6}) exposed people is accepted (a 'political' and/or policy decision). These are calculated by linear extrapolation to zero exposure (the spontaneous incidence), which in principle is a worst-case approach because no threshold is assumed.

mutagens' and virtually all cells are subjected to spontaneous mutations and 'naturally' occurring DNA damage.

This issue is further complicated if there is an effect of 'hormesis' for genotoxic agents. The 'hormesis' theory states that a chemical that is known to be toxic at higher doses may be associated with a beneficial effect at low doses. For example, there are some data suggesting that a low level of exposure to a genotoxin stimulates the activity of the high-fidelity DNA repair, and it has also been shown that low levels of dioxin consistently decrease the incidence of certain tumours in animal experiments.

As shown in Figure 1.7, essential elements such as nutrients and vitamins (which also are toxic at high doses) have a rather unusual dose–response curve when it comes to their adverse health effects. A deficiency results in malfunction, while an excess results in toxicity.

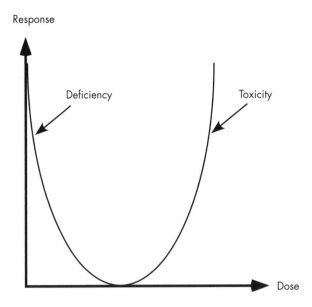

Figure 1.7 The toxicity response of an essential element. Typically a U-shaped dose–response curve (theoretical) is expected for an essential element such as selenium or oxygen: when the supply is too low a deficiency leads to disease, while an excess leads to toxicity.

Toxicity testing is necessary for hazard identification and safety assessment

In toxicology, a distinction is made between hazard and risk. Whereas *hazard* is an 'inherent' potential of a chemical to be toxic at some site (e.g. it has the potential to cause malformations in rabbits or cancer in the forestomach of rats), *risk* includes the determination of the chance of the expression of an identified hazard given the exposure characteristics of the chemical. The concept of risk therefore includes two components: probability and outcome (in toxicology, an adverse health effect). Consequently, whereas risk is the probability that a compound will produce harm under specific conditions, safety (the reciprocal of risk) is the probability that harm will not occur under the specified conditions. The concept of risk is typically used for non-drugs; that of safety for drugs. Completely different, but very important, issues when discussing the risk/safety of chemicals are the processes of risk communication, risk perception and risk acceptance. For example, whether a risk is voluntary or not, is known or unknown, is of natural origin or human-made, etc., plays a decisive role in the degree of acceptance of a risk.

For example, a patient will most likely accept rather serious side-effects if the disease to be treated is directly life-threatening or debilitating. It is interesting to note that agents like botulinum toxin (BTX; an exotoxin with extremely high acute toxicity), arsenic trioxide (As_2O_3, 'the king of poisons') and thalidomide (a well-known human teratogen; see Chapter 4) are used today as drugs to treat various diseases, such as cervical dystonia (BTX), acute promyeloid leukaemia (As_2O_3) and hypertrophic cutaneous lupus erythematosus (thalidomide).

A major purpose of toxicity testing is hazard identification, that is to identify the 'toxicological profile' of a drug (or any chemical) and characterise its 'inherent' potential to act as a toxicant. By adding data on toxicokinetics in animal studies, as well as pharmacokinetic data for the patient when treated with the intended dose, it is also possible to calculate a margin of safety for a drug (see below). The spectrum of adverse health effects ('toxicological end-points') that can be induced by a chemical includes both reversible and irreversible effects, local and systemic toxicity, immediate and delayed effects, and organ-specific and general adverse effects.

In combination with toxicokinetic studies, toxicity testing provides information about the shape of the dose–effect curves for the various types of toxic effects identified, including approximate lethal doses in acute toxicity studies, TD_{50} values as well as NOAELs and LOAELs (see Figures 1.4 and 1.5). The LD_{50}, that is the dose that kills 50% of the treated animals, is no longer used for drugs for reasons of saving the animals from unnecessary suffering; in any case, such values are not relevant for clinical side-effects of drugs in patients. The TD_{50} is the dose that induces an adverse effect in 50% of the exposed individuals.

The most valuable information about the toxicity of chemicals derives from observations made in exposed humans, but deliberate toxicity testing on human subjects is, of course, out of the question. Of necessity, toxicity data are therefore generated in studies on experimental animals. The results obtained in these studies are then extrapolated to the human exposure situation. The whole concept of toxicity testing is based on the presumption that experimental animals can be used to identify potential health hazards for humans. For drugs, the situation is usually much simpler: here human data become available during clinical trials. It is essential that all drug metabolites formed in humans also are present in the toxicity studies on animals, since every metabolite is expected to have its own unique toxicological profile.

As shown in Table 1.1, toxicity tests can be separated into two major categories: those designed to identify general toxicity (typically,

Table 1.1 Toxicological end-points measured in an extensive battery of toxicity tests

Adverse health effect	A representative sample of tests for which there are internationally accepted testing guidelines
Acute toxicity and lethality	• Acute oral toxicity study in rats • Acute inhalation toxicity study in mice • Acute dermal toxicity study in guinea pigs
Short-term repeat dose and subchronic toxicity	• 14-day oral toxicity study in mice • 21-day dermal toxicity study in rats • 90-day diet study in rats • 90-day inhalation toxicity study in guinea pigs
Chronic toxicity	• 12-month oral toxicity study in dogs • 24-month inhalation toxicity study in rats
Genotoxicity and mutagenicity	• *Salmonella typhimurium* reverse mutation assay • Mouse lymphoma TK-locus assay • Micronucleus test
Carcinogenicity	• 18-month dietary cancer bioassay in mice • 24-month inhalation cancer bioassay in rats
Reproductive toxicity	• Two-generation reproduction study in rats
Embryotoxicity and teratogenicity	• Teratogenicity study in rabbits • Teratogenicity study in rats
Allergic sensitisation	• Skin sensitisation study in guinea pigs
Local effects on the skin	• Acute dermal irritation and corrosion study in rabbits
Neurotoxicity	(No specific testing guidelines available for neurotoxicity or behavioural toxicity; indirect evidence in general toxicity studies)
Immunotoxicity	(No specific testing guidelines available; thymus toxicity can be monitored in general toxicity studies)

toxic effects in various organs assessed by, for example, histopathology or clinical chemistry parameters), and those designed to identify specific types of adverse health effects (e.g. genotoxicity, cancer, reproductive toxicity and teratogenic effects). The general pharmacology studies on drugs also include what often is called 'safety pharmacology', focusing on effects of the drug on vital physiological functions such as those of the heart, the lungs and the central nervous system. In some cases, special studies may be required when a certain toxic effect of a drug in animals raises concern and the drug company wants to prove that this effect is not relevant for the proposed human use. Nowadays drugs should also be evaluated with regard to their environmental safety (using different types of ecotoxicological studies). This need arises from concern about, for example, ethinylestradiol (a component of oral contraceptive pills) in waste water, and its oestrogenic effects in the environment.

For most toxicity tests there are internationally accepted guidelines describing how each individual test should be performed so as to obtain a well-defined framework for both the testing and the evaluation of toxicity; for instance, the OECD (Organisation for Economic Co-operation and Development) Guidelines for Testing of Chemicals. Much information about toxicity testing can be found on OECD's home page (www.oecd.org). The various guidelines specify the prerequisites, procedures (preferred species, group sizes, duration of treatment, etc.), testing conditions (dosages, routes of administration, clinical observations, haematology, pathology, etc.), statistical procedures, and how the test report should be formulated. The guidelines have also been developed in order to minimise suffering of experimental animals. Protection of the animals from unnecessary stress due to pain and discomfort is important not only from a humane point of view but also because stress may interfere with the 'true' toxic responses induced by the chemical under test.

It should be emphasised that all toxicity tests are designed to reveal potential toxicity and not to prove the harmlessness of chemicals. The tests are therefore usually designed to be as sensitive as possible. For ethical, practical and economic reasons, toxicity testing is usually performed using a restricted number of animals. Relatively high doses are used (for statistical reasons) in order to compensate for the limited number of animals. 'High' doses do not mean lethal doses. For example, the highest dose in a chronic cancer bioassay is often referred to as the 'maximum tolerated dose' (MTD). This dose should be high enough to induce some signs of toxicity (e.g. a slightly reduced body weight gain), but it should not substantially alter the normal lifespan of the animals as a result of effects other than tumour development.

The possible shortcomings of toxicity testing on animals and cultured cells must always be considered when evaluating the results of these tests. Data from *in vivo* studies in animals allow calculations of margins of safety but may not necessarily be predictive of the ultimately relevant adverse effects in the patients. Primates (especially great apes) are more comparable to humans than rats, for example, but the use of primates is limited because of societal concern. Animal toxicity studies could, for example, predict that a drug will induce kidney toxicity in patients, but it might very well turn out later in clinical studies that this drug actually induces certain unacceptable CNS effects, and that it is the latter effects rather than the kidney toxicity that limit its human use.

Consequently, clinical side-effects observed in Phase II and Phase III studies will ultimately determine human toxicity profiles under therapeutic conditions. Unfortunately, rare side-effects may still be missed. Even in relatively large Phase III trials employing, say, 3000 patients, rare side-effects with an incidence of less than 1 in 1000 (risk level = 1 $\times$ 10^{-3}) will almost certainly be missed unless they are extremely unusual (in the general population). This is why post-marketing surveillance and periodic safety updates ('PSUR') are extremely important after introducing a new drug to the market.

Several factors determine which (if any) side-effects will be observed in the general patient population at the therapeutic dosage. Among these are ethnicity, age, body weight, health condition, nutritional status, lifestyle (alcohol, smoking habits and drug abuse), co-medication, duration and frequency of exposure and/or climate conditions. Since ethnic differences can be quite pronounced in some cases, clinical studies in different ethnic groups are required. Many of the considerations mentioned above are also important when evaluating animal toxicity data, because they can influence the outcome in a toxicity study (along with species, strain, sex and route of exposure). Despite all the possible limitations of animal toxicity studies and the undeniable fact that different animal species may respond differently to various toxicants, there is a general consensus that animal studies are required and useful (but not sufficient) to guarantee the safe use of drugs.

In vitro assays can be used to study mechanisms and identify (potential) hazards, but they do not give information about the magnitude of risk.

Margins of safety

A safe drug has a large margin of safety (MOS). At the group level (but not necessarily on the individual level) this means that unwanted side-effects of such a drug require a much higher dose than is required for the therapeutic effect. The MOS can be calculated in different ways depending on the type of chemical and the available data. If the MOS is based on toxicokinetic data for the internal exposure in experimental animals and on pharmacokinetic data for the internal exposure in humans, one will get a rather reliable estimate of the MOS. For drugs, these data (blood levels, etc.) are usually readily available from both the toxicokinetic and toxicity studies in animals and the clinical trials.

In the past, a MOS for drugs was usually calculated as the ratio of the dose that is just within the lethal range (LD_1) to the dose required to result in a therapeutic response in 99% of the exposed patients (ED_{99}). Clearly, LD_1 (the dose that kills 1% of exposed animals) and ED_{99} were statistically derived (e.g. by extrapolating from probit-transformed data) and not measured directly (an impossible task). The ratio LD_1/ED_{99} has at least one advantage over the therapeutic index ($TI = LD_{50}/ED_{50}$), another statistically derived approximate estimate of the relative safety of drugs. Since TI is based on median values (see Figure 1.4), it only compares the mid-points of the dose–response curves and does not consider at all the slopes of the dose–response curves for the therapeutic and toxic effects. If there is an overlap in these two curves, a sensitive patient might experience adverse effects without the therapeutic effect, even if the average patient is effectively treated (without adverse effects).

Numerical values of LD_1/ED_{99} and LD_{50}/ED_{50} are no longer calculated; instead, margins of safety for drugs are based on toxicokinetic parameters that reflect the internal exposure much better. Consequently, MOS values for drugs are usually based on a toxic effect in animals (e.g. liver necrosis) occurring at a certain blood concentration and the blood concentrations reached in the patient on a therapeutic dosage schedule. If the difference is big enough, MOS is judged acceptable, but, obviously, the size of an acceptable (or unacceptable) MOS is to a large extent dependent on the severity of the disease to be treated.

For residues of pesticides in food, environmental pollutants in the air, and other types of non-drugs (for which there are no beneficial effective therapeutic doses), margins of safety (sometimes referred to as margins of exposure; MOE) can be calculated by comparing the difference between NOAEL (typically obtained in a toxicity study on animals) and the estimated exposure level in the human population at risk.

Studies on general toxicity

Acute toxicity

The main purpose of acute toxicity testing of drugs is to gather basic toxicological information before the more extensive toxicity testing and

Phase I clinical trials in healthy human volunteers (see Chapter 13). After exposure of a restricted number of animals (orally, intravenously or by inhalation), these are examined at least once a day for 14 days (including clinical chemistry in blood and urine). Animals showing severe signs of intoxication are killed prematurely to spare unnecessary suffering, and all animals are subjected to an autopsy at the end of the study.

The acute toxicity studies not only provide important information on immediate health hazards and clinical signs of intoxication; they can also identify possible target organs and conceivable modes of action. The results obtained in acute toxicity studies are used when establishing the dosage regimens in repeated dose toxicity studies. For chemicals in general, the results can also be used to classify them in terms of various types of toxicity ratings (see Table 1.2 and 1.3). Depending on the acute toxicity data, including its estimated lethality, a chemical can be classified as being 'very toxic', 'toxic', 'harmful' or 'moderately harmful' using a classification system based on LD_{50} or LC_{50} values in

Table 1.2 Examples of single risk phrases for compounds that should be classified as very toxic (symbol T+) if after a single, brief exposure they can cause temporary or irreversible injuries or lead to death. (Taken from the Swedish National Chemicals Inspectorate)

Code	Phrase	Criterion
R28	Very toxic if swallowed	• LD_{50} oral, rat $\leq$ 25 mg/kg • Less than 100% survival at 5 mg/kg (oral rat) according to the discriminating dose (the fixed dose procedure)
R27	Very toxic in contact with skin	• LD_{50} dermal, rat or rabbit $\leq$ 50 mg/kg
R26	Very toxic by inhalation	• LC_{50} inhalation, rat, aerosols and particulates $\leq$ 0.25 mg/L, 4 h • LC_{50} inhalation, rat, gases and vapours $\leq$ 0.5 mg/L, 4 h
R39	Danger of very serious irreversible effects	• Strong evidence that irreversible injuries can occur, generally within the above-mentioned dose ranges, after a single, brief exposure via a relevant route of administration. Note: Carcinogenicity, mutagenicity and reproductive toxicity excluded

Table 1.3 The connection between classification in categories of danger and labelling: some examples. (Taken from the Swedish National Chemicals Inspectorate)

Category of danger	Symbol letter	Symbol
Very toxic (R26, R27, R28 and R39)	T+	Skull and crossbones
Toxic (R23, R24, R25, R39, R48)	T	Skull and crossbones
Corrosive (R34, R35)	C	Corrosion symbol
Harmful (R20, R21, R22, R65, R40, R48)	X_n	St Andrew's cross

experimental animals (see Table 1.4), or as 'super toxic' (<5 mg/kg), 'extremely toxic' (5–50 mg/kg), 'very toxic' (50–500 mg/kg), 'moderately toxic' (0.5–5 g/kg), 'slightly toxic' (5–15 g/kg) or 'practically nontoxic' (>15 g/kg) based on its estimated oral lethal dose for humans (adults). Many recommendations regarding protective measures and possible need for medical attention when handling chemicals are based on such acute toxicity classification systems.

Short-term repeated dose and subchronic toxicity

Short-term repeated dose and subchronic toxicity studies provide information on immediate and delayed adverse effects, possible bioaccumulation, reversibility of damage, and development of tolerance. The clinical and histopathological examinations are quite extensive and it is therefore often possible to establish NOAELs and LOAELs that can be

Table 1.4 Classification categories based on LD_{50} or LC_{50} values. (Taken from the Swedish National Chemicals Inspectorate)

Classification category	Acute toxicity	LD_{50} (oral) (mg/kg, rat)	LD_{50} (dermal) (mg/kg, rat)	LC_{50} (inhalation) (mg/kg, rat)
Very toxic	Very high	≤25	≤50	≤0.5[a] (≤0.25)[b]
Toxic	High	25–100	50–400	0.5–2[a] (0.25–1)[b]
Harmful	Medium high	200–2000	400–2000	2–20[a] (1–5)[b]
Moderately harmful	Moderate	>2000[c]	—	—

[a] Applies to gases and vapours.
[b] Applies to aerosols and particulates.
[c] Applies to consumer products.

used when establishing the dosage regimen in other toxicity studies. For short-term human drug use (days or weeks), this information can be used to assess the safety/efficacy balance. Moreover, for short-term occupational exposures this type of information can also be used when establishing a threshold limit value and other types of safety criteria.

Chronic toxicity

The main purpose of chronic toxicity testing is to identify adverse health effects that require long periods to develop, and to establish dose–response relationships for these effects. The chronic toxicity studies can provide important information on various types of bio-chemical, haematological, morphological, and physiological effects. Using rodents, each dose group (including a concurrent control group) should include at least 20 animals of each sex. When non-rodents are used (less commonly), a minimum of four animals (usually dogs or primates) of each sex is recommended per dose. The test compound should be administered on a daily basis in at least three different doses for at least 12 months. For drugs, toxicokinetic data should be collected at regular time intervals in so called 'satellite groups' that are treated for that specific purpose only.

A careful check should be made daily, and all clinical signs of toxicity (changes in body weights, food consumption, behaviour, etc.) should be recorded. Measurements of blood clinical chemistry, haematological examinations and urinalysis should be performed on a regular basis, and all animals should be subjected to a complete gross examination at necropsy. The histopathology should include at least a complete micro-scopic examination of most of the organs and tissues from the animals in the highest dose group and the control group, including all animals that died or were killed during the study period. The NOAELs and/or LOAELs obtained in a chronic toxicity study are typically used to establish a margin of safety for drugs or to establish various threshold limit values for environmental and/or occupational human exposures.

Studies on various types of specific adverse health effects

Genotoxicity

Genotoxicity (genetic toxicity) represents a diversity of genetic end-points including primary DNA damage (DNA adducts, cross-linking, intercalation, DNA strand breaks, etc.), gene mutations (changes in the nucleotide sequence at one or a few coding segments within a gene), structural chromosomal aberrations (e.g. translocations and inversions following chromosome or chromatid breakages), and numerical chromosomal aberrations (e.g. aneuploidy and other genomic mutations). There are numerous test systems available to detect various types of genetic end-points, using a broad spectrum of 'indicator' organisms (from plants, bacteria, yeast, insects, and cultured mammalian cells, to intact experimental animals). So far there are internationally accepted testing guidelines for approximately 15–20 of these test systems. The main purpose of genotoxicity testing is to establish whether a given compound has the 'inherent' ability of being mutagenic (usually with the aim of identifying potential carcinogens). For a more detailed survey of genetic toxicology, see Chapter 5.

Carcinogenicity

Tumour development is a multistage process involving both permanent genetic alterations (i.e. mutations) and other ('epigenetic') events. The neoplasms ('tumours', 'cancers', 'malignancies') are a family of diseases characterised by aberrant control of cell proliferation and cell differentiation, and most malignant diseases are multifactorial in origin. The main purpose of a chronic cancer bioassay is to study the potential development of tumours in experimental animals exposed for a major portion of their lifespan. Typical indications of carcinogenicity are development of types of neoplasms not observed in controls; increased incidence of types of neoplasms also observed in controls; occurrence of neoplasms earlier than in the controls; and/or increased multiplicity of neoplasms in animals exposed to the test compound. Chronic cancer bioassays are usually performed on two different species (typically mice and rats), using at least 50 males and 50 females per dosage for each species (including an unexposed group of control animals). For a more detailed survey of chemical carcinogenesis, see Chapter 6.

Reproductive toxicity

Toxic responses of the reproductive system can be the result of disruption of spermatogenesis or oogenesis (gametogenesis), or of adverse effects on libido, fertilisation, implantation, embryogenesis, organogenesis, fetal growth, or postnatal development. In a broad sense, reproductive toxicity studies include single- and multi-generation studies on fertility and general reproductive performance (segment I studies), studies on embryotoxicity and teratogenicity (segment II studies; see below), and peri- and postnatal studies on effects occurring during late gestation and lactation (segment III studies).

The main purpose of a typical one- or two-generation reproductive toxicity study is to provide information on chemically induced adverse effects on the male and female reproductive performance. By studying parturition, duration of gestation, number and sex of pups, stillbirths, live births, microscopic alterations in the gonads of the adult animals, or gross anomalies in the offspring, for example, information can be obtained on adverse effects on gonadal function, oestrous cycle, mating behaviour, conception, parturition, lactation and weaning. These studies should also be able to provide some information on developmental toxicity, including neonatal morbidity and behaviour. In a typical segment I study, both sexes (usually rats) are exposed to graduated doses of the test compound (in order to cover important stages of both male and female gametogenesis). After mating, the females are continuously exposed during gestation and the nursing period as well. In a two-generation study, the test compound is also given to the offspring (the F_1 generation), starting at weaning and continuing until the second generation (the F_2 generation) is weaned. Some mechanisms behind reproductive effects are discussed in Chapter 4.

Embryotoxicity, teratogenicity and fetotoxicity

Chemicals may affect the developing embryo or fetus without inducing any overt signs of maternal toxicity. For example, depending on the stage of embryonic or fetal development, a toxicant may induce early embryonic death, fetal death, malformations, retarded maturation or low birth weight, as well as various metabolic and physiological dysfunctions and cognitive deficiencies in the offspring. In the broadest sense, embryotoxicity can be defined as all types of adverse effects exhibited by the embryo (i.e. toxicity occurring from the formation of the blastula until the completion of organogenesis). Fetotoxicity (i.e. the adverse effects exhibited by the fetus) is induced after the completion of organogenesis. Typical examples of such effects are increased lethality at birth, low birth weight, various types of physiological and psychological dysfunction, and cognitive disturbances manifested after birth.

To be able to distinguish between the various types of adverse health effects that may follow from exposure to a chemical during gestation, 'embryotoxicity' has also come to mean the ability of a chemical to impair embryonic growth or induce embryonic death. Teratogenicity is typified by permanent structural or functional abnormalities (including external malformations, skeletal abnormalities and/or visceral anomalies), but may also include behavioural changes (behavioural teratology) if a wider definition is used.

The main purpose of teratogenicity testing is to provide information on the potential hazards to the unborn following exposure during pregnancy. Embryotoxicity and teratogenicity without apparent maternal toxicity are particularly alarming. The test compound is given in different doses to pregnant animals (usually rats or rabbits), for a period including organogenesis. As in most other toxicity studies, the highest dose administered should elicit some maternal toxicity, and the lowest dose should be without apparent signs of toxicity. The pregnant animals are killed shortly before the expected time of delivery, and the offspring are examined for various embryotoxic and teratogenic effects. For a more detailed description of mechanisms behind teratogenic effects, see Chapter 4.

Neurotoxicity and behavioural toxicity

Neurotoxicity can be defined as chemically induced adverse effects on any aspect of the central and peripheral nervous system, including various supportive structures. From this it follows that 'neurotoxicity' is associated with various types of pathological changes in the nervous system that are expressed as changes in morphology, physiology, biochemistry and/or neurochemistry, as well as various types of functional and neurobehavioural changes. Obviously, neurotoxicity is not a single end-point that can be evaluated in a single test system. Pathological changes in various regions of the brain, and/or clinical signs of intoxication deriving from CNS toxicity (e.g. piloerection, tremor or coma) can be monitored in acute and repeated dose toxicity studies. Chemically induced behavioural changes (sometimes very subtle) are more difficult to monitor. This usually requires a completely different type of testing procedure, with which a 'traditionally' trained toxicologist is not always familiar.

Test systems are available to detect various types of (subtle) CNS effects (e.g. changes in reflexive or schedule-controlled behaviours, or reduced performance in various learning and memory tasks). The concept of behavioural toxicology is based on the notion that a behaviour is the final functional expression of the whole nervous system (indirectly including the endocrine system and other organs as well) and that behavioural changes therefore can be used as sensitive indicators of chemically induced neurotoxicity, both in adult animals and in animals exposed *in utero* or shortly after birth ('neurobehavioural teratology').

Behavioural toxicity tests are based on changes either in spontaneous behaviour of the animals (e.g. their natural social or exploratory behaviour) or in stimulus-oriented behaviour. The latter tests are either directed towards an operant-conditioned behaviour (the animals are trained to perform a task in order to avoid a punishment, or to obtain a reward) or towards classical conditioning (the animals are taught to associate a conditioning stimulus with a reflex action). Typical responses recorded in the various types of behavioural toxicity tests are 'passive avoidance', 'auditory startle', 'residential maze' and 'walking patterns'. It is not always easy to interpret the results from behavioural neurotoxicity tests in terms of (relevance to) human behaviour. Apart from the obvious problems associated with the functional reserve and adaptation of the nervous system, there is also an inherently large variability in behaviour. Since neurobehavioural testing usually involves multiple measurements in several different test systems, there is an obvious risk of getting some statistically significant result purely by chance.

Immunotoxicity including sensitisation

Toxic effects mediated by the immune system are sometimes referred to as immunotoxicity. However, from a toxicological point of view, immunotoxicity is in most cases defined as chemically induced adverse effects on the immune system. The immune system is a highly complex and cooperative system of cells, tissues and organs, protecting the organism from infections and neoplastic alterations. Immunotoxic agents can interact with the immune system in many ways. They can, for example, interfere with the function, production or lifespan of the B- and T-lymphocytes, or interact with various antigens, antibodies and immunoglobulins. Basically, immunotoxic agents function either as 'immunosuppressive' agents or as 'immunostimulants'. Consequently, like several other toxicological 'end-points', immunotoxicity is not a single end-point that can be monitored with a single test system.

Skin sensitisation, i.e. allergic contact dermatitis, is an immunologically mediated reaction requiring an initial contact with an agent that induces sensitisation. It can be difficult to distinguish between an allergic contact dermatitis caused by skin sensitisation and an 'ordinary' contact dermatitis following skin irritation, because the symptoms (typically involving erythema and oedema, sometimes also including vesicles) are quite similar. However, when sensitisation has occurred, responses are often more severe and are also elicited by rather low, non-irritating doses without apparent thresholds. Being very sensitive, the guinea pig is the preferred species when testing for skin sensitisation.

Several alternative tests are available: for example, Freund's complete adjuvant test, the guinea pig maximisation test and the open epicutaneous test. Most of these tests follow the same general outline. To make the animals hypersensitive, they are first exposed to a rather high dose of the test compound. After an 'induction period' without exposure, the animals are then exposed a second time to a low, non-irritating dose. After this challenge dose, the animals are examined with regard to possible development of allergic contact dermatitis. Sensitisation can also occur after other routes of exposure (notably inhalation), but internationally accepted testing guidelines have so far been developed only for allergic contact dermatitis.

Several drugs (e.g. penicillins and halothane) cause immune reactions ('allergy') because they can function as haptens. Since this type of allergic reaction is very difficult to predict in animal models, drug-related side-effects of this type are usually not detected until after the clinical trials or after the drug has been on the market for some time. Hypersensitivity responses to drugs are actually among the major types of unpredictable drug reactions and the effects can be drastic, including anaphylactic shock in the case of penicillins and hepatitis in the case of halothane. For a more detailed survey of immunotoxicology, see Chapter 10.

Skin and eye irritation

The main purpose of testing for local toxicity on skin and eyes is to establish whether a chemical induces irritation (reversible changes) or corrosion (irreversible tissue damage) when applied as a single dose on the skin or to the anterior surface of the eye. Obviously, there is no point in testing a strongly acidic (pH ≤ 2) or alkaline (pH ≥ 11.5) agent for local toxicity on the skin and eyes. Equally, if an agent has been shown to be corrosive to the skin, it seems rather pointless to proceed with an acute eye irritation study. Testing for local effects on the skin is usually performed on albino rabbits (each animal serving as its own control). The degree of skin reaction is read and scored at various time points (up to 14 days after the application). Depending on the degree of erythema and oedema on the skin, the test chemical is classified as a non-irritant, irritant or corrosive agent. Currently, eye irritation tests are often done *in vitro* rather than in the traditional Draize test on rabbits' eyes *in vivo*.

Toxicodynamics: there are many different mechanisms of toxicity

In parallel with the concept of pharmacodynamics, one may summarise the adverse health effects of a drug and the mechanisms behind them as 'toxicodynamics'. Most toxicants induce their adverse effects by interacting with normal cellular processes, and many toxic responses are the ultimate result of cell death leading to loss of important organ functions. Other responses follow from interactions with various biochemical and physiological processes not affecting the survival of the cells. Common mechanisms of toxic action include receptor–ligand interactions, interference with cellular membrane functions, disturbed calcium homeostasis, disrupted cellular energy production, and reversible or irreversible binding to various proteins, nucleic acids and other 'biomolecules'. Toxicity can be the result of one specific physiological change in a single target organ, or can follow from multiple interactions at different sites in several organs and tissues.

Many toxicants induce their adverse effects by binding to a specific site on a biologically active molecule. This molecule can be a protein (e.g. a 'high-affinity' receptor, a bioactivating or detoxifying enzyme, a DNA-repair enzyme, a channel protein or a transport protein), a nucleic acid (DNA or RNA), a lipid, or another macromolecule with important biological functions. A 'receptor' is usually defined as a high-affinity binding site interacting with an endogenous ligand. Typical examples of such receptors are those interacting with various neurotransmitters in the CNS, and the intracellular receptors interacting with, for example, calcium or various steroid hormones. However, in a broad sense a receptor can be defined as any binding site available for a particular ligand, and in that sense the definition of a 'receptor' is broader in toxicology than in pharmacology.

When a toxicant binds to a high-affinity receptor for an endogenous ligand, it can either 'activate' the biological responses mediated by the receptor (acting as an 'agonist'), or block its function (acting as an 'antagonist'). The agonist can act directly by binding to the receptor or indirectly by increasing the concentration of the endogenous ligand at the receptor (e.g. by inhibiting its degradation, as with acetylcholinesterase inhibitors).

There are numerous examples of toxicants acting by binding to various macromolecules. For example, the anoxia resulting from the high-affinity binding between carbon monoxide and haemoglobin is an example of an adverse effect that is due to binding to a protein, in this case non-covalent binding. 'Metabolic poisons' interfere with the biological activity of various enzymes. Some toxicants do this by binding to the enzymes and thereby changing their structure. Other types of metabolic poisons interfere with the metabolic pathways by competitive inhibition. Toxicants can also interfere with cellular energy production, for instance by inhibiting oxidative phosphorylation in the mitochondria. Such agents (e.g. several anti-AIDS drugs) are usually called 'mitochondrial poisons'. Other toxicants act as 'cellular poisons' by interfering with various membrane-bound functions and transport processes. Among those are many potent neurotoxins acting as ion channel blockers by binding to various channel proteins.

Several drugs form reactive intermediates during their biotransformation. These electrophilic intermediates can bind directly to various cellular macromolecules, but they can also induce 'oxidative stress' in the cells. This will eventually lead to the formation of various reactive oxygen species, including highly reactive hydroxyl radicals interacting with, for example, DNA (causing DNA damage) and unsaturated fatty acids in the cell membrane (causing lipid peroxidation). Oxidative stress has been implicated as an important factor in many biological processes, including ageing, inflammatory reactions and tumour development. Lipid peroxidation has been implicated as a mechanism of action for many hepatotoxic agents inducing centrilobular liver necrosis. For a more detailed survey of mechanisms behind cytotoxicity, see Chapter 3.

The evaluation of toxicity data is not always straightforward

Most toxicological data are derived from animal experiments. Toxicity studies identify the nature of health damage that may be associated with a given compound, and the range of doses over which the damage is produced. When such data are used for safety assessment (of drugs intended for medical use) or risk assessment (of contaminants in occupational and environmental settings), the methods used in the toxicological evaluation should always be critically assessed. Was the toxicant administered accurately? Were the means of expressing the toxicity precise? Did the study include an adequate number of animals and dosages? Did the study include a control group? Were the results statistically significant? Were the responses biologically significant?

The results from one type of study should always be interpreted in conjunction with the results obtained in other toxicity studies and one should also be aware of the limitations when extrapolating animal data to a human exposure situation. For chemicals regulated in the occupational and environmental area, so-called uncertainty (or safety) factors are often used when extrapolating animal data to humans. Typically a NOAEL obtained in a toxicity study in animals is divided by a total uncertainty factor of 100 (10×10). One factor of 10 is to compensate for the possible interspecies variability (animals to humans) in toxicokinetics and toxicodynamics, the other factor of 10 is to compensate for the intraspecies variability (human to human). However, the uncertainty factors used may vary depending on the situation. For example, a factor of 3 may be used for a more homogenous worker population (instead of 10 for the general population), and additional safety factors may be added if the toxicity data are based on LOAELs only, or if the adverse health effect is considered to be particularly serious (e.g. a malignant disease).

For drugs intended for medical use in humans, an exceptionally wide range of toxicity data is available because of the requirements for registration: both animal and human data in relation to the intended therapeutic use and route of administration. The latter requirement means that new data have to be generated when a new route of administration of is considered for a drug that is already on the market. Moreover, if a drug has a registered indication for a short-term use (e.g. maximum three weeks for short-term treatment of pain) and the drug company wants an indication for chronic use (for example, to treat arthritic pain), this will require carcinogenicity data that were not needed for the short-term treatment.

The evaluation of toxicological data is often rather straightforward for immediate adverse health effects following from a well-characterised chemical exposure. Nevertheless, we still often do not know what to do during severe drug overdose because the mechanism behind the acute toxicity is unknown (unless it is a pharmacological side-effect). For continuous low-dose exposures (less of a problem for drugs, but a well-known issue in occupational and environmental medicine), the toxicological evaluation can become complicated (for example, involving questions of whether there is a threshold or not).

A complicating factor when evaluating toxicity data is the development of tolerance – a decreased responsiveness towards the toxicity of a chemical resulting from a previous exposure. Tolerance is a process of adaptation to the effects of a compound, which becomes obvious when, for example, it is necessary to increase the dosage in order to obtain a given therapeutic response. Tolerance may be due to downregulation of receptors or to selection of resistant cells (not uncommon for cytotoxic cytostatics). Tolerance to morphine and ethanol are two well-known examples of adaptation. Similarly, initial irritation in the nose may disappear after an exposure period of a few minutes.

Another complicating factor is that chemicals, including drugs, can interact with each other. This issue is not usually addressed in conventional toxicity testing of chemicals, which focuses on one compound at a time. The interaction between chemicals can result in an additive effect (the combined effect equals the sum of the effect of each individual agent given alone); a synergistic effect (the combined effect is much greater than the sum of the effect of each agent given alone); a potentiating effect (one of the agents is without toxic effects of its own to the particular organ system, but when given together with a toxic agent, it will multiply the toxic effect); or an antagonistic effect (the combined effect is lower than the sum of the effects of each agent given alone).

During drug development, the company must address the possibility of interaction problems in relation to the drug's intended use, including both pharmacokinetic and pharmacodynamic interactions, as well as toxicokinetic and toxicodynamic interactions. For example, will the drug in question inhibit certain cytochrome P450 enzymes that are needed for the elimination of other drugs? Or is it possible that co-medication with other drugs might inhibit the metabolism of the new drug? Drugs that are frequently used by the intended patient population have to be tested with regard to their pharmacodynamic interactions.

Conclusion

When a company submits the dossier of a certain indication for the therapeutic use of a new drug, it has to provide exhaustive evidence for the safety of that drug in comparisn with its efficacy and intended use, taking into consideration the seriousness of the disease to be treated. Depending on the duration, the route of administration and the dose, the registration requires a certain set of data. Only when the registration authorities are convinced that the balance between safety and efficacy is positive will the drug be registered for that indication. For other, non-drug chemicals, much less is usually known about their toxicological profiles (especially as regards toxicokinetics and pharmacodynamics in humans). In such cases, experimental data from *in vivo* studies on animals and/or *in vitro* assays on isolated cells have to be substituted for human data (making the conclusions about potential risks less certain). Only for bulk chemicals that have been in use for many years may enough epidemiological data be available to allow conclusions about their human toxicity.

The fact that most toxicity data are still generated in experimental studies (on animals and/or cells) makes it imperative that any professional working with different aspects related to the safe use of chemicals, including drugs intended for medical use, should understand fundamental toxicological principles and have a basic knowledge of the premises for toxicity testing.

Further reading

Klaasen CD, ed. (2001). *Casarett & Doull's Toxicology. The Basic Science of Poisons*, 6th edn. New York: McGraw-Hill.

Hayes AW, ed. (2001). *Principles and Methods of Toxicology*, 4th edn. Philadelphia: Taylor and Francis.

2

Drug metabolism: inactivation and bioactivation of xenobiotics

Gerard J Mulder

General introduction

Some history

In the 19th century the first experiments were done to find out what happens to chemicals ('xenobiotics': exogenous chemicals in the body) when they are taken up in the body. In one case, an assistant drank some benzene and the investigator collected urine to see whether certain new components could be found with the analytical tools available at that time. Thus various 'urates' were identified: *mercapturates* (urinary compounds containing sulfur) and *hippurates* (metabolites discovered in horse urine). This work continued in the 20th century, leading in 1947 to the famous book *Detoxification Mechanisms* by R.T. Williams. This book conveyed the message that toxic chemicals were converted in the body to non-toxic metabolites. Quite complex, species-specific metabolite patterns were discovered for a wide variety of chemicals.

Subsequently, in the 1960s and 1970s, several groups showed that biotransformation could also lead to more toxic metabolites, in particular reactive intermediates that were responsible for carcinogenesis (if they formed DNA adducts) or for cell death. The presence of these reactive intermediates had to be inferred because most were too short-lived for their formation to be demonstrated directly. Work by the Millers, Brodie and Gillette revealed the chemical mechanisms behind this reactivity and the toxic action. On the basis of this work, Ames developed the *Samonella* 'Ames' test for mutagenicity of chemicals and their metabolites. The role of cytochrome P450 isozymes in the bioactivation of chemicals to reactive intermediates was defined. The conjugation (Phase 2) reactions still retained their image of mediating detoxification, although it soon became quite clear that these could also be responsible for toxification.

In this chapter only the major routes of drug metabolism for 'classical' drugs will be briefly discussed and examples of the consequences for efficacy and safety of drugs will be given. The increasingly developed therapeutic proteins (e.g. humanised antibodies or insulin analogues) are broken down by proteolytic enzymes that are also used for the elimination of endogenous proteins.

Why do we have drug-metabolising enzymes? First-pass metabolism

In their natural environment, animals – including humans – are exposed to many chemicals that might pose a danger if they accumulate in the body: food from animal and plant sources contains many chemicals of no nutritional value but of potential toxicity. Similarly, the air may contain such chemicals (in some cases indicated by their smell!). If they are sufficiently lipid-soluble they will reach the blood; for the same reason, they will not readily be excreted unless they are converted to more water-soluble metabolites. This may be the reason why all animals have a wide variety of drug-metabolising enzymes that convert a wide range of chemical structures to water-soluble metabolites that can be excreted in urine (or bile). The high activity of many of these enzymes in the gut mucosa and the liver ensures that internal exposure to many such chemicals is limited: a high percentage of the absorbed dose may be caught in the first pass after entering into the body, so that systemic exposure is quite limited. At the same time, this so-called *first-pass effect* limits the availability of medicines after oral administration, because these compounds are converted by the same (drug-metabolising) enzymes. Moreover, the first-pass effect may contribute to hepatotoxicity, because of the high uptake (= exposure) and bioactivation activity in the liver.

Family members of most of the drug-metabolising enzymes are also essential for the metabolism of endogenous compounds, both to eliminate them and to generate metabolites with a physiological role.

General characteristics of drug-metabolising enzymes

Classically, drug metabolism has been divided into two phases (Figure 2.1). In Phase 1 metabolism the chemical structure of the substrate is modified by oxidation, reduction or hydrolysis. In many cases this leads to formation of an acceptor group for Phase 2 metabolism – conjugations – whereby a chemical group is attached to the acceptor group. In

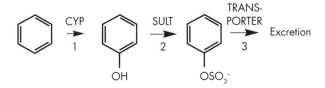

Figure 2.1 The three phases of drug metabolism, illustrated for benzene. CYP = cytochrome P450; SULT = sulfotransferase.

general this results in much more water-soluble metabolites that are more readily excreted than the parent compounds. The transport of a drug or its metabolites out of the cell catalysed by transporters is often defined as Phase 3 metabolism.

A very wide range of (usually related) enzymes or enzyme families catalyses each of the major biotransformations. They may overlap in substrate specificity so that the same substrate can be converted by more than one form of the enzyme. The following applies to most of the drug-metabolising systems:

- The drug-metabolising enzymes form *super*families consisting of many families and subfamilies of more or less related isoforms.
- The nomenclature contains up to four terms, e.g. CYP2A4, defining superfamily (CYP), family (2), subfamily (A) and isoform (4). Information about mutant forms can be added at the end (e.g. CYP2A4*4). The species is defined by *m* for mouse, *h* for human, etc. Orthologues (i.e. 'the same enzyme-form') between species can be identified, but they may not be completely identical in, say, substrate specificity.
- Several genes (possibly on different chromosomes) code for the enzymes within the same superfamily.
- Gene splicing occurs frequently so that several isoforms can be 'written' from the same gene.
- In each of these enzymes there occur mutations (point mutations or deletions) that may result in changes in the activity (including complete loss of function).
- Gene multiplication may lead to increased expression of an enzyme in certain individuals. For instance, up to 13 copies of CYP2D6 are present in some individuals (the 'very extensive metaboliser' phenotype). Therefore, interindividual differences between humans are to be expected due to gene (expression) differences and mutations.

- The expression of the enzymes can increase on exposure to chemicals ('induction'). This may be quite specific or non-specific; for instance, many chemicals induce both CYP and UGT forms simultaneously. Transcriptional suppression is also possible but has been much less studied.
- The individual isoenzymes are expressed at (very) different levels in the various organs and species. This may give rise to organ-specific or species-specific toxicity. During fetal development, some enzymes may be present early, others only after birth. In general the levels of drug-metabolising enzymes are rather low in fetuses.

Table 2.1 summarises the most common biotransformation reactions for a number of functional groups.

Table 2.1 Common biotransformations of functional groups

Functional group	Enzymes involved (product)
Alcoholic –OH, –SH	ADH, UGT, SULT, MT
Aldehyde	Aldehyde dehydrogenase
Phenolic –OH	UGT, SULT, MT
Amine	CYP (hydroxylamine, etc.), FMO (N-oxide), acetylT, UGT, SULT (sulfamate), MT
Hydroxylamine	CYP (nitroso- and nitro-derivative), UGT, SULT
Epoxide	EH, GST (addition)
Carboxylate	UGT (acyl-glucuronide), amino acid conjugation
Halogenated compounds	GST (nucleophilic displacement)
Double bond	CYP (epoxide), GST (addition)
Aromatic ring	CYP (phenol or epoxide)

UGT, UDP-glucuronosyltransferase; UDP, uridine diphosphate; SULT, sulfotransferase; MT, methyltransferase; ADH, alcohol dehydrogenase; GST, glutathione S-transferase; CYP, cytochrome P450; FMO, flavine monooxygenase; acetylT, acetyltransferase; EH, epoxide hydrolase.

Phase 1 metabolism

Oxidative metabolism: cytochrome P450 (CYP)

The major drug-metabolising pathway is catalysed by cytochrome P450 enzymes (known in the older literature as mixed-function oxidases). The abbreviated notation is CYP. Over 60 human CYPs have been identified. CYP3A4 and CYP2D6 are of special importance, because they convert a great variety of drugs. CYP3A4 and CYP3A5 represent approximately 50% of total hepatic CYP and are responsible for approximately half of the drugs metabolised by CYPs. By far the most

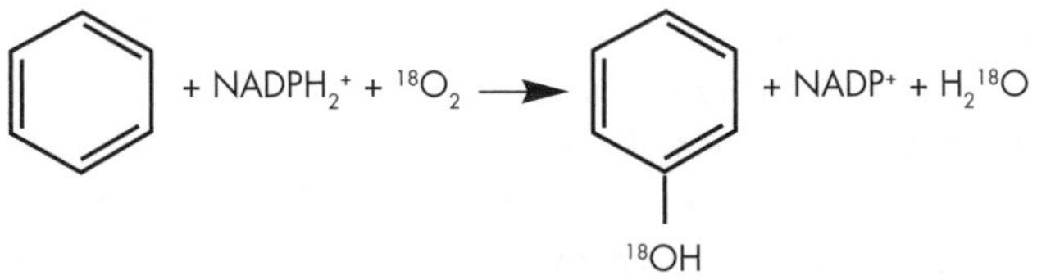

Figure 2.2 CYP reaction for benzene, showing the destination of the oxygen atoms derived from molecular oxygen labelled with ^{18}O.

CYPs are located in the endoplasmic reticulum and are recovered from tissue homogenate in the microsomal fraction. Some, however, are located in mitochondria for example. The basic reaction catalysed by CYP is shown in Figure 2.2. The fate of the two oxygen atoms is indicated by using ^{18}O molecular oxygen ($^{18}O_2$): one atom is incorporated into water and the other into the substrate (as can be detected by mass spectrometry).

The characteristic feature of CYP enzymes is the haem group, which binds oxygen such that it accepts electrons from NADPH (through a reductase). The oxygen molecule is split into two oxygen atoms. One of these is converted to water, the other attacks the acceptor substrate, resulting in an oxidised metabolite. A mechanistic model of the CYP reaction is shown in Figure 2.3. The cycle starts by binding of the substrate to the CYP species with iron in the Fe(III) state (step 1). Next an electron is added by CYP reductase, which converts the iron to Fe(II) (step 2), followed by binding of an oxygen molecule and transfer of the electron to the oxygen. At this stage, under certain unfavourable conditions, a molecule of superoxide anion, $O_2{}^-$, may be released, leading to toxicity (e.g. DNA damage, or cytotoxicity). Usually, however, an additional electron is added (step 4) and two protons are used to form a water molecule from the first oxygen atom. The high-valent iron-oxo complex finally attacks the substrate, oxidising it in the final step.

Because in many cases a C–H bond has to be broken by CYP action, replacement of the hydrogen atom by a deuterium atom (and even more so by a tritium atom) results in slower oxidation: the C–D or C–T bond is stronger than the C–H bond. This is the *deuterium effect* or *tritium effect*. If a fluoride atom replaces H in a C–H bond, that position becomes very resistant to metabolism, which is a reason to have fluorine atoms in some drugs. In some cases deuterium or even methyl groups can shift to a different position in a substrate molecule during

CYP attack, the so called *NIH-shift* (which was discovered at the National *I*nstitutes of *H*ealth in Bethesda, MD, USA).

Many CYPs are involved in the conversion of endogenous compounds, thus fulfilling a role in physiological processes. For instance, thromboxane synthetase and prostacyclin synthetase are members of CYP5A and CYP8A families, respectively. The CYP11 family includes enzymes catalysing steroid hydroxylation. Such CYPs are much more substrate-specific than those for xenobiotics.

Typical examples of CYP-mediated oxidative metabolic routes are shown in Table 2.2. CYP may also be involved in reductive reactions, an important example being the reductive metabolism of halothane (see p. 63). In that case an electron is transferred by CYP to the substrate without subsequent oxidation.

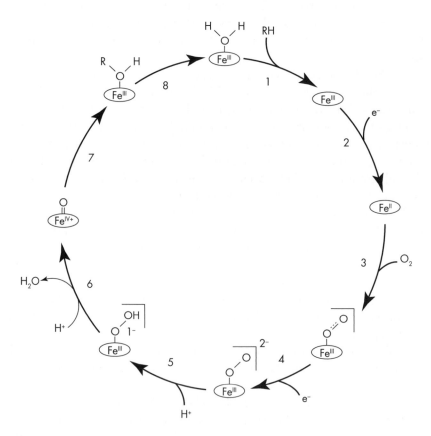

Figure 2.3 The CYP cycle. (Modified from Makris, Davydov, Denisov, Hoffmann, Sligar (2002) Mechanistic enzymology of oxygen activation by the cytochromes P450. *Drug Metab Rev* 34: 691–708.)

Table 2.2 Some typical CYP-mediated reactions

Type of reaction	Substrates	Products
Aromatic hydroxylation	Aromatics	Phenol, catechol
Aliphatic hydroxylation	Alkanes	Alcohol
Epoxidation	Aromatics, unsaturated aliphatics	Epoxides
O-, S- or N-dealkylation	N-, S- or O-alkyl groups	Aldehydes
N-oxidation	Amines	Hydroxylamines
S-oxidation	Diverse S-derivatives	Sulfoxides
Oxidative and reductive dehalogenation	Halogen-containing alkanes	Various

As a result of CYP action, the substrates usually undergo a change in their pharmacological activity: some compounds become more active, others less active than the parent. A pronounced increase in toxicity or pharmacological activity may be the result ('bioactivation', see pp. 61–65).

CYP inhibitors

CYP inhibitors may be active towards a number of CYPs, or more specific for a single CYP. *In vitro*, carbon monoxide is an effective inhibitor because it blocks the haem group. Of particular importance are the reversible inhibitors, among which are many drugs. These inhibit by competition for the substrate-binding site or by non-competitive mechanisms. Examples of such inhibitors are given in Table 2.3. Some substrates are suicide inhibitors of CYPs: their conversion results in reactive intermediates that bind covalently to the components in the active site and thus irreversibly block the enzyme.

Several highly efficient *in vivo* inhibitors of CYPs are available. For instance, SKF 525A has been widely used to confirm the role of CYP in the metabolism of a compound *in vivo*. Drugs that inhibit CYPs cause (sometimes fatal!) interactions (see p. 65). Some foods, such as grapefruit juice, also contain inhibitory components. The potential inhibitory action of a new chemical entity (NCE) is tested with CYP isoenzymes *in vitro*. The extent of inhibition *in vitro* need not be the same as that *in vivo*, where it may be much higher but in some cases can be lower.

Table 2.3 Some substrates, inhibitors and inducers of CYP

Enzyme	Substrate/inhibitor/inducer	
CYP1A2	Substrates:	paracetamol, caffeine, imipramine, warfarin
	Inhibitors:	furafylline, a-naphthoflavone
	Inducers:	omeprazole, cigarette smoke
CYP2C9	Substrates:	phenytoin, piroxicam, tolbutamide, (S)-warfarin
	Inhibitors:	sulfaphenazole, sulfinpyrazone
	Inducer:	rifampicin
CYP2D6	Substrates:	captopril, fluoxetine, imipramine, mianserin, paroxetine
	Inhibitors:	ajmalicine, quinidine, trifluperidol, yohimbine
CYP3A4	Substrates:	budesonide, ciclosporin, diazepam, nifedipine
	Inhibitors:	clotrimazole, ketoconazole, naringenin
	Inducers:	carbamazepine, phenobarbital, phenytoin, rifampicin

Induction

Induction by numerous chemicals is another feature of CYPs; most of the CYPs are inducible. Several receptor-mediated mechanisms converge in the cell nucleus, where responsive elements effect increased gene transcription, e.g. xenobiotic- and glucocorticoid-responsive elements (XRE and GRE respectively) driven by ligand-activated transcription factors. The increased mRNAs are subsequently translated into increased levels of the CYPs. Which CYP species is induced depends on the inducer. A number of inducers with their specificity of induction are listed in Table 2.3.

The biological significance of induction appears to be adaptation to a 'chemical stress': increased exposure to the chemical results in an increased capacity to eliminate it. For drugs, this implies that the dose has to be increased to keep the effect constant.

Both inhibition and induction may lead to potentially dangerous drug–drug or drug–diet interactions in patients.

Several drugs have been used as probes in humans to evaluate CYP activity *in vivo*. Thus midazolam is metabolised mainly by CYP3A4 to two hydroxylated metabolites. Phenozone (antipyrine) is metabolised by demethylation by a number of CYPs; its metabolism is quite sensitive to inducers and inhibitors. Such probes can be used to evaluate potential interaction with other chemicals *in vivo*.

Many CYP polymorphisms have been described; these are briefly discussed in the section Polymorphisms and Genetic Variation below.

Oxidative metabolism: flavin-containing monooxygenases (FMO)

These are microsomal enzymes that contain a flavin group and use NADPH as well as oxygen to oxidise, in particular, compounds containing nitrogen, sulfur or phosphorus heteroatoms. Examples are formation of an *S*-oxide from thioureas or cimetidine (Figure 2.4), and *N*-oxidation of imipramine or chlorpromazine. Because these enzymes are not induced by any of the known enzyme inducers, and also inhibition by other chemicals is rather inefficient, substrates of FMO may show fewer interaction problems than those of CYPs.

A deficiency in FMO3 causes affected individuals to have a very unpleasant odour. FMO3 rapidly metabolises the odorous trimethylamine to its *N*-oxide. However, this does not occur in FMO3-deficient persons, leading to accumulation of trimethylamine, which results in an offensive fish odour.

Oxidative metabolism: monoamine oxidase (MAO)

A limited range of substrates is metabolised by monoamine oxidases (MAO). These enzymes are involved in the metabolism of many endogenous biogenic amines such as adrenaline (epinephrine) or dopamine. Figure 2.5 shows a typical MAO conversion. The enzymes are flavoproteins and use NADH as electron donor. They are located in mitochondria. The reaction products are an aldehyde and H_2O_2.

Because depression was believed to be due to a deficiency in certain biogenic amines, irreversible MAO inhibitors have been used as antidepressants. Although they were effective, they had major side-effects because they irreversibly inhibit an essential enzyme. Interaction with

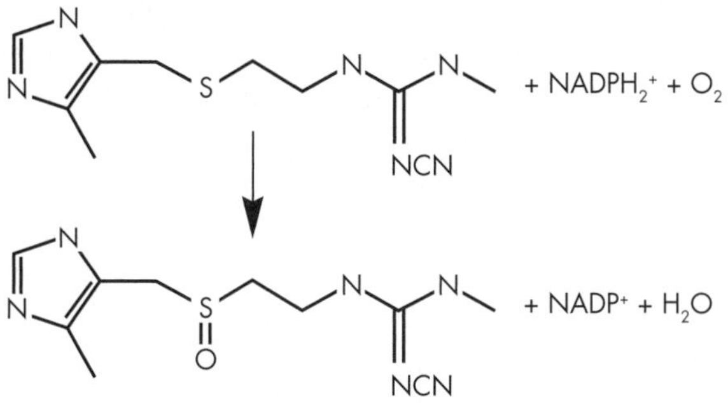

Figure 2.4 FMO-catalysed formation of an *S*-oxide from cimetidine.

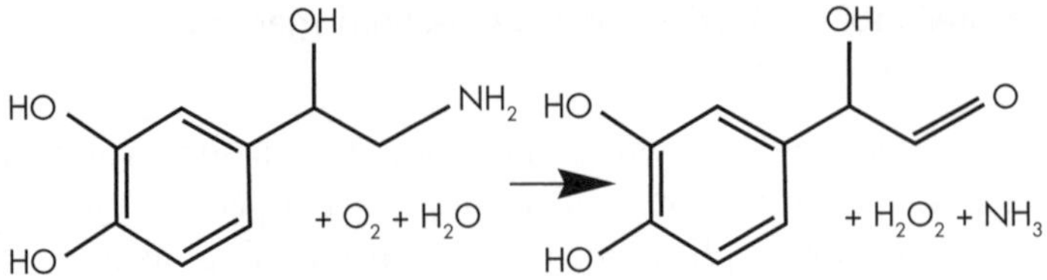

Figure 2.5 MAO-catalysed oxidation of noradrenaline (norepinephrine).

amines, in particular tyramine, in food (cheese, certain wines) resulted in fatal accidents: the MAO inhibitor also inhibits tyramine metabolism, so that tyramine concentrations accumulate. In patients with a 'weak heart', usually elderly people, this resulted in a lethal effect.

In addition, MAO activity can be responsible for bioactivation. Thus, MPTP, a designer drug, is converted by MAO to the neurotoxin MPP+ in astrocytes; MPP+ subsequently destroys the action of dopaminergic neurons (Figure 2.6).

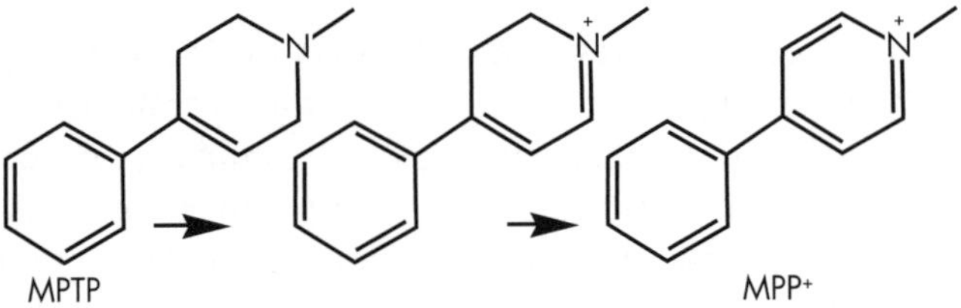

Figure 2.6 Bioactivation of MPTP to a neurotoxic metabolite MPP+.

Hydrolysis

Esters and amides can be hydrolysed by carboxylesterases. The enzymes require only water as co-substrate. These enzymes belong to a superfamily of proteins. The highest activity of many carboxylesterases is in the liver microsomal fraction.

The enzymes may be rather specific (e.g. acetylcholinesterase) or non-specific. Non-specificity is utilised in the case of many ester-type prodrugs: since the ester group shields a hydrophilic carboxyl or hydroxyl group, it increases the lipid solubility of a compound so that its absorption from the intestine is increased. Esterification may also lead to a prolongation of half-life of the active metabolite because of its slower generation by hydrolysis from the prodrug. Esterases hydrolyse

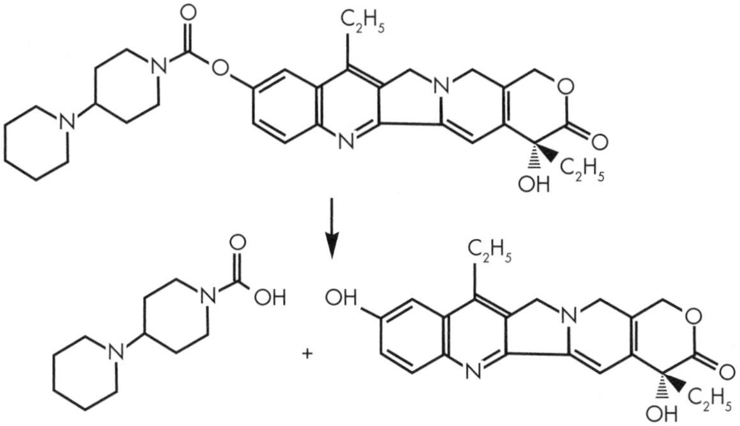

Figure 2.7 Hydrolysis of irinotecan.

such prodrugs, which then release the active substance. An example is irinotecan, a topoisomerase I inhibitor, which is rapidly converted by carboxylesterases to its active metabolite (Figure 2.7). Some patients have low-activity polymorphic variants of the carboxylesterases, which may be why irinotecan is not effective in these patients.

Acetylcholinesterase is essential for many physiological functions such as muscular action and CNS activities. Compounds that inhibit its activity may therefore have strong biological actions. This feature has been abused to develop organophosphorus chemical warfare agents that bind irreversibly to a serine moiety of the acetylcholinesterase active site (Figure 2.8). This is lethal upon sufficient (inhalatory) exposure. The same effect is caused by some insecticides such as parathion. The inhibitors form a covalent adduct with a serine in the active site of the enzyme. By administration of a nucleophilic agent such as pralidoxime,

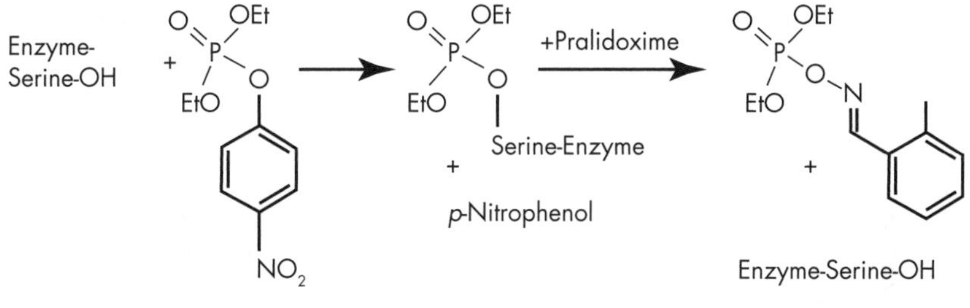

Figure 2.8 Acetylcholinesterase inhibition by paraoxon and reactivation by pralidoxime.

the adduct can be removed from the serine residue, which reactivates the enzyme and prevents toxicity if it is administered very promptly after exposure to the agent.

Some reversible acetylcholinesterase inhibitors have been introduced for clinical use against Alzheimer's disease; presumably acetylcholine deficiency plays a role in these patients and it is claimed that these inhibitors have a beneficial effect in delaying the progression of Alzheimer's disease in some patients.

Reduction

Reductases, especially in gut bacteria, may reduce chemicals in the intestinal lumen. In mammalian tissues also, reductive reactions play a role, including some reactions catalysed by CYP and other cytochromes. For instance, (aromatic) nitro compounds can be reduced to the nitroso, hydroxylamine or amine stages (Figure 2.9). This is of toxicological significance because the hydroxylamines may be involved in carcinogenic action: they may form DNA adducts.

Another example is the reduction of a quinone to the semiquinone, eventually followed by generation of superoxide anion if the electron is subsequently transferred to oxygen. This may give rise to redox cycling (Figure 2.10), a potential source of cytotoxicity: the quinone, the semiquinone and the superoxide anion generated may cause severe toxicity to the cell.

Finally, CYP may catalyse reductive dehalogenation, e.g. of halothane (see p. 63).

Epoxide hydrolase

Epoxide hydrolases (EHs) catalyse the addition of a water molecule across an epoxide bond (Figure 2.11). The result is a dihydrodiol, which is generated in a stereospecific manner. Although this inactivates a

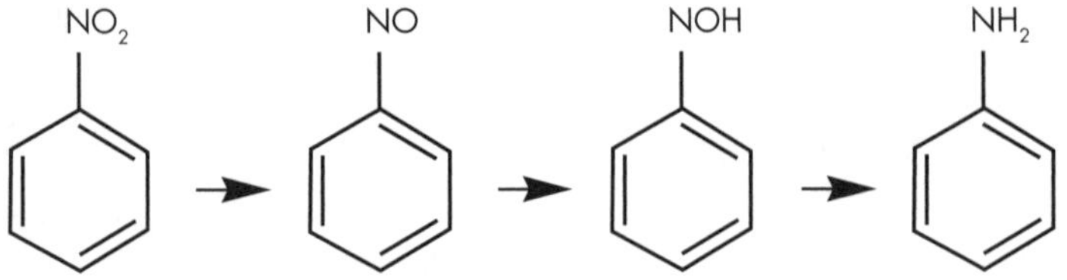

Figure 2.9 Nitroreductase reduction of nitrobenzene.

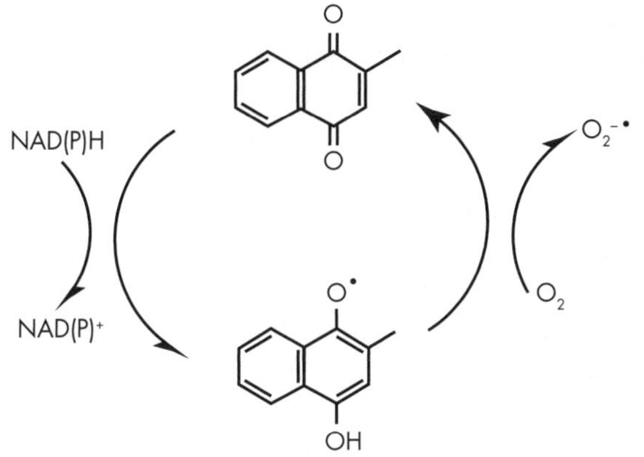

Figure 2.10 Redox cycling.

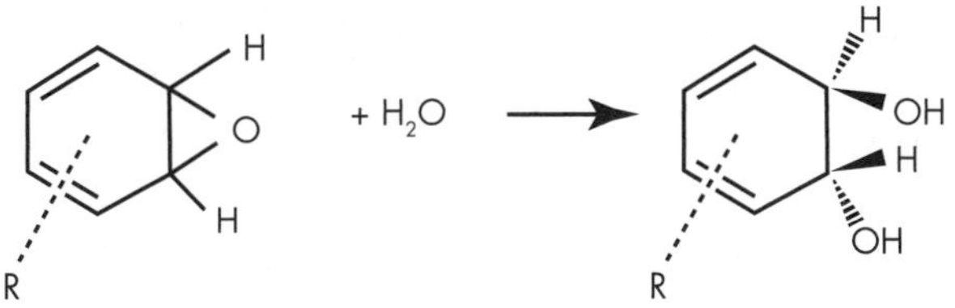

Figure 2.11 Epoxide hydrolase reaction.

potentially reactive epoxide, it may also be a step on the way to an ultimate carcinogen, as in the case of benzo[a]pyrene. Physiological epoxides are also substrates for EHs, in particular lipid epoxides that may have functions in blood flow or inflammation, for example. Both cytosolic (soluble, sEH) and membrane bound (mEH) forms are found.

Phase 2 reactions (conjugations)

Co-factor availability

Conjugation reactions are catalysed by transferases. They ultimately have to make a covalent bond between, in most cases, a nucleophilic group in the acceptor substrate and the conjugating group. This requires energy. In most conjugations this is provided by the co-substrate,which is an 'activated' form of the group to be transferred (e.g. PAPS (phosphoadenosine phosphosulfate), UDPGA (UDP-glucuronate), SAM (S-adenosylmethionine), acetyl-CoA or the acyl-CoA for amino acid conjugation). In the case of glutathione (GSH) the conjugating enzyme uses the (nucleophilic) thiol of GSH in an activated form to attack the already electrophilic atom in the acceptor substrate.

Table 2.4 lists the most common co-factors as well as the requirements for their biosynthesis.

Conjugations require the transfer of a group from the co-factor to the acceptor substrate; for instance, the transfer of the glucuronate group from UDPGA to a phenolic group. As a consequence the availability of the co-factor may become limiting in conjugation. At peak demand, the availability of every co-factor can in principle be (temporarily) decreased: for GSM, sulfate (actually 'active sulfate' or PAPS) and the amino acid conjugations there are clear indications that co-factor depletion may play a significant role. Thus, in paracetamol toxicity, depletion of detoxifying GSH precedes cell toxicity (see p. 62). Depletion of sulfate (and therefore of sulfation) may lead to a compensatory increase of glucuronidation of the same substrate. This may have toxicological consequences if the sulfate conjugate has a different toxicity from the glucuronide. For instance, the sulfate conjugate of N-hydroxy-2-acetylaminofluorene (N-OH-2AAF) is extremely reactive and carcinogenic, whereas the corresponding glucuronide conjugate is not. Clearly in this case it makes a lot of difference which of the pathways is followed!

Dietary factors determine whether co-substrate depletion may occur. Thus, a diet low in protein may lead to a shortage of cysteine and derived products such as GSH and sulfate.

Table 2.4 gives an overview of the conjugations that are *in principle* possible for a number of functional groups; the actual contribution of a conjugation reaction depends on the overall structure of the

Table 2.4 Co-factors for major conjugations

Conjugation	Co-factor	Biosynthesis requirements
Glucuronidation	UDP-glucuronate (UDPGA)	UTP + glucose 1-phosphate
Sulfation	3'-phospho-5'-adenosine phosphosulfate (PAPS)	$2ATP + SO_4$
Glutathione conjugation	Glutathione (Gly-Cys-γGlu) (GSH)	Amino acids + 2ATP
Methylation	S-Adenosylmethionine (SAM)	ATP + methionine
Acetylation	Acetyl-CoA	ATP
Amino acid conjugation	Acyl-CoA derivative	ATP

compound as well as on neighbouring groups that may cause steric hindrance or have special electronic effects.

Glucuronidation

Glucuronidation of substrates takes place at a nucleophilic group in the acceptor substrate and is catalysed by the UDP-glucuronosyltransferases (UGTs). The UGTs are microsomal enzymes with the special property that they are 'latent' in microsomal preparations. A detergent or similar membrane-active compound has to be added to produce maximum activity, often many times times greater than the 'native' activity. A wide variety of xenobiotics with phenolic or carboxylic groups are eliminated by glucuronidation (Figure 2.12). More than 16 human UGTs have been identified. Some of these isoforms occur only extrahepatically, e.g. in nasal epithelium or the gastrointestinal tract, probably in relation to their physiological function. Indeed, some of the UGTs play a major role in the metabolism of endogenous compounds. For instance, the elimination of bilirubin requires its glucuronidation (by UGT1A1), and many steroids are in part excreted as glucuronides. UGT2B7 glucuronidates steroids and retinols as well as morphine, zidovudine (azidothymidine, AZT), nonsteroidal anti-inflammatory drugs (NSAIDs) and valproate. Highly selective *in vivo* inhibitors have not yet been found.

Many UGTs can be induced by the same compounds as those inducing CYPs. The UGTs utilise UDP-glucuronic acid as group-donating

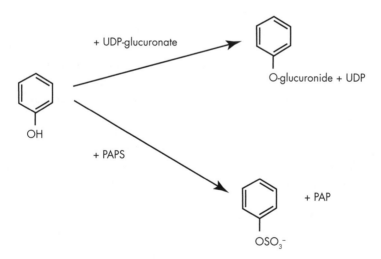

Figure 2.12 Glucuronidation and sulfation reactions of phenol.

co-substrate. Since a large glucuronate group is attached to the acceptor substrate, the biological properties of the products are usually considerably altered because, for example, they no longer fit in the binding site of the receptor for the parent compound. In some cases a glucuronide is still biologically active, like the 6-glucuronide of morphine. Only rarely are they toxicologically relevant, such as steroid D-ring glucuronides, which are cholestatic. However, acyl-glucuronides, formed from carboxyl groups, are in general reactive and may be the cause of idiosyncratic reactions (see p. 64).

The UGTs show pronounced polymorphisms, of which the clinical consequences are so far poorly defined. A human disease, Gilbert's syndrome, is characterised by jaundice, which is due to a hereditary deficiency of UGT1A1. In these patients the elimination of some drug substrates is also affected, although there is compensation by other UGT family members.

Several other sugars can also be utilised for conjugation – for instance, glycosides are generated from UDP-glucose by UDP-glycosyl-transferases – but these are less common than UGT products.

Sulfation

The sulfation reaction is catalysed by the sulfotransferases (SULTs). Substrates contain a nucleophilic group (phenols, alcohols), which is conjugated with a sulfonate group. SULT1A1 conjugates drugs such as paracetamol and minoxidil but also endogenous 3,5,3'-triiodothyronine. SULT1E1 is highly specific for 17β-estradiol in the nanomolar range. In addition, many SULT forms are involved in sulfation of, for example, sugar moieties of glycosaminoglycan or tyrosine residues in proteins. These cytosolic enzymes are not readily induced by other chemicals, but glucocorticoids and tamoxifen can induce SULT1A1. Effective *in vivo* inhibitors for various types of SULT are available, such as pentachlorophenol.

The SULTs use PAPS as a group-donating co-substrate. For phenols and alcohols in humans, sulfation is usually the preferred reaction; at higher doses, glucuronidation may prevail. Thus, thyroxine in humans is sulfated rather than glucuronidated, while in the rat glucuronidation is the preferred reaction.

In general, sulfates are much less biologically active than their parents, and thus detoxification appears to be the rule. However, the active form of the antihypertensive minoxidil is its sulfate conjugate, and many steroid hormones are active in their sulfated form. It is possible that the biological activity of many sulfates has been overlooked because often they have not been tested, on the assumption that sulfates simply are not active. Toxicologically, sulfation plays a major role in the bioactivation of hydroxamic acids and benzylic alcohols to reactive intermediates that may form DNA adducts and thus cause cancer. Tamoxifen provides an example of bioactivation to a carcinogenic metabolite (see p. 63).

The sulfation reaction requires PAPS. In mice, a deficiency in PAPS synthesis is known, resulting in mice with very short legs ('brachymorphic'), probably because sulfated components of bone tissue cannot be synthesised. The defect is due to deficient biosynthesis of sulfate-containing bone components and also has consequences for sulfation of xenobiotics.

Acetylation

The acetyltransferases (ATs) catalyse the transfer of the acetyl group from acetyl-CoA to the acceptor group, which in most is cases an amine (Figure 2.13) but can also be an oxygen atom of a phenol or an alcohol. The enzymes are located in the cytosol. Acetylation was the first drug metabolism for which polymorphism was observed, as well as pronounced ethnic differences.

No selective *in vivo* inhibitors of acetylation are available. Induction of acetyltransferases has not been observed so far.

Amino acid conjugation

A compound with a carboxyl group can be converted to a CoA derivative that can subsequently be conjugated to an amine – usually glycine or taurine, but in certain species other amines such as ornithine are also used (Figure 2.14). If conjugation to glycine occurs, the so called 'hippurates' are formed. Bile salts are commonly such conjugates of cholate to taurine or glycine. These conjugates are efficiently excreted in urine or bile. No induction has been reported for the enzymes involved, nor are selective *in vivo* inhibitors known.

Figure 2.13 Acetylation of aniline.

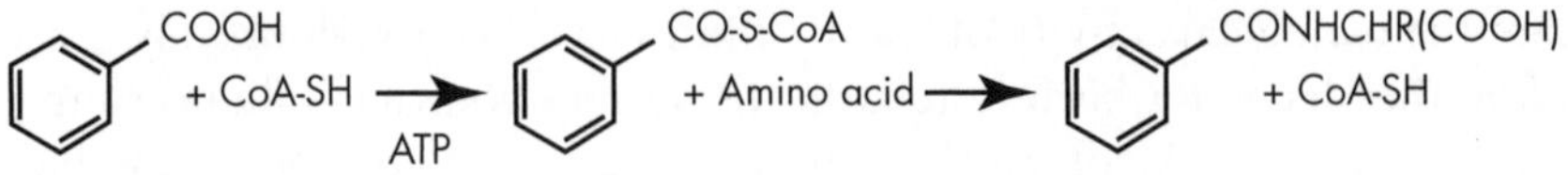

Figure 2.14 Amino acid conjugation.

Glutathione conjugation

Substrates with a (strong) electrophilic atom in their structure are subject to glutathione (GSH) conjugation catalysed by the glutathione S-transferases (GSTs). The enzymes are in almost every case cytosolic, but microsomal forms are also known. They function prominently in detoxification of reactive intermediates such as epoxides and quinones (including those generated from endogenous compounds), but are also involved in the formation of certain prostaglandin derivatives with physiological functions. The GSTs can be induced by many chemicals; this may be considered an adaptive response since many chemicals are detoxified by GSTs (Figure 2.15). No highly specific *in vivo* GST inhibitors are available. Inhibition of GSH conjugation is usually achieved by depletion of GSH with diethylmaleate or phorone. These treatments are very non-specific because they will affect every

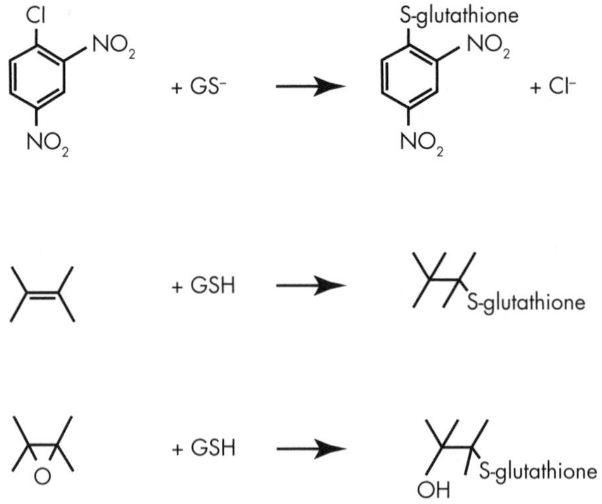

Figure 2.15 Glutathione (GSH) conjugation of various substrates: nucleophilic displacement of chlorine, addition of GSH to a carbon–carbon double bond, and nucleophilic attack at an electrophilic carbon atom in an epoxide.

GSH-dependent process. Many physiological processes depend on GSH and, therefore, its depletion may easily lead to cell toxicity.

Because the electrophilic substrates are made less reactive, in general GSH conjugation implies detoxification. However, vicinal halogens like 1,2-dibromoethane are converted to reactive, genotoxic glutathione conjugates.

The GSH conjugates are further metabolised, predominantly in the kidney, to mercapturates (meaning 'sulfur-containing compounds in urine') (Figure 2.16). Usually these are non-toxic, but in some cases they may give rise to highly nephrotoxic thiols, in particular from halogenated alkanes or alkenes.

Deficiencies in glutathione synthesis have been observed. Several enzymes in the γ-glutamyl cycle can be decreased. These patients are very rare, suggesting that in most cases such deficiencies will be lethal and lead to early abortion. Generalised GSH deficiency obviously will have very serious consequences for all GSH-dependent processes.

Methylation

The methyltransferases catalyse the transfer of the methyl group from *S*-adenosylmethionine to, in most cases, an amine or a (phenolic) hydroxyl group (Figure 2.17). The enzymes are catechol *O*-methyl transferases or *N*-methyltransferases. They are not readily induced by chemicals, and no selective *in vivo* inhibitors are available. Some metals can be methylated, e.g. mercury (Hg^{2+}) or arsenic. This may have important consequences for the toxic effects: methylation makes the metal ion more lipid-soluble so that it can reach sites that are not easily accessed by the unconjugated metal ion, such as the brain.

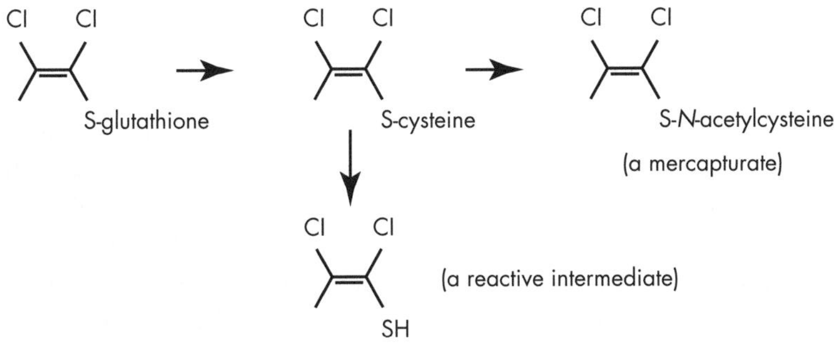

Figure 2.16 Mercapturate pathway and toxification of halogenated compounds.

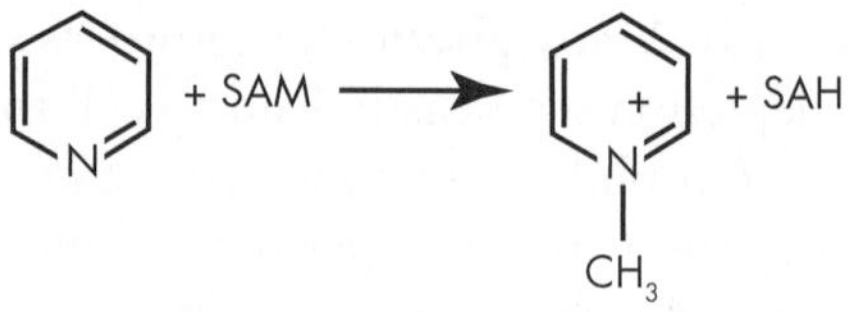

Figure 2.17 Methylation of pyridine. SAM = S-adenosylmethionine; SAH = S-adenosylhomocysteine.

Choice of animal species for drug development

One of the first things a pharmaceutical company does when it considers taking a chemical from discovery to development is to find out how the compound is metabolised in humans as well as in animal species to be used for safety assessment, with particular attention to the formation of reactive intermediates. If a particular animal species is to serve as a model for assessment of toxicity to be expected in humans, it should represent a similar spectrum of metabolites as that in the human patient. Each metabolite has its own characteristic toxicity profile and therefore all of them need to be included in the safety assessment of a prospective drug. A comparable metabolite pattern is thus obligatory in safety studies. In some cases it is necessary to synthesise a metabolite that is specific for humans if it is not a metabolite in any of the common animal species. For instance, quaternary ammonium glucuronides are formed only in humans and great apes. These metabolites, therefore, have to be synthesised chemically and evaluated for their toxicity in rats or mice by direct administration of the compounds.

The concentration of major metabolites in the blood in animal studies is followed to allow a toxicokinetic evaluation of exposure in comparison with the human exposure at therapeutic doses.

Stereoisomers

Drug metabolism is enzyme-mediated; therefore, stereoselectivity towards enantiomeric substrates is observed. Similarly, when a chiral centre is generated by metabolism, the product is often a single stereoisomer. The metabolism of styrene offers a nice example. Styrene 7,8-oxide is the primary CYP metabolite; depending on the species (and the CYPs involved), the R- or S-isomer predominates, the R-isomer being somewhat more toxic. Subsequently, epoxide hydrolase and GST may further metabolise the oxide to a wide variety of chiral metabolites. The final outcome in terms of metabolites and their stereochemistry is strongly species- and dose-dependent (see *Drug Metabolism Reviews* (2003) 33: 353). Since the pharmacological activity as well as toxicity of stereoisomers may be (very) different, such stereoselectivity in drug metabolism may have important consequences.

Polymorphisms and genetic variation

A wide variety of polymorphisms has been discovered for the drug-metabolising enzymes, and many of these are relevant for drugs in terms of efficacy and safety. Certain isoenzymes may be deficient because they are not present in an active form or are not present at all. Point mutations may lead to forms that are less active than the most common one in the population. This applies to all biotransformations. Single nucleotide polymorphism (SNP) analysis has resulted in many entries in databanks collecting the nucleotide sequences for the different (iso)enzymes.

Many attempts have been made to link certain diseases or even processes such as ageing to such SNPs. However, the possibility for the organism to functionally compensate a deficiency in one (iso)enzyme by other forms of the enzyme or by compensatory pathways (e.g. sulfation and glucuronidation are mutually compensatory to a certain extent) will 'soften' many of the consequences. Animals in which an (iso)enzyme is 'knocked out' serve as useful models for such diseases. Often the consequences become observable only under stress conditions, e.g. when an animal is challenged with a certain toxic drug, because the phenotype may otherwise not show abnormalities.

Such polymorphic variations will not only affect the metabolism of drugs but may also have major consequences for the endogenous substrates, thus leading to diseases.

In the case of the conjugations, not only the transferases are relevant but also the enzymes responsible for the synthesis of the co-substrates. Examples include defects in PAPS synthesis (brachymorphic mice) or GSH biosynthesis (several human diseases).

Bioactivation to toxic metabolites and idiosyncratic reactions

Biotransformation can give rise to toxic metabolites in a bioactivation process. For many drugs the mechanisms behind bioactivation and ensuing toxicity have been unravelled at the cellular and molecular levels. However, there are unanticipated and as yet unexplained toxic side-effects of drugs. It is likely that some of these 'idiosyncratic reactions' are due to the formation of reactive intermediates that lead to an immune reaction to the drug moiety in some organ. For instance, if the reactive intermediate is formed in the liver it may form a protein adduct (a 'hapten') and reach the outer membrane of the liver cell. Attack by the immune system may lead to cell death and ultimately hepatitis, as is the case for halothane. Other mechanisms will also contribute to idiosyncratic events. Here a few examples will be given of toxic effects with a widely accepted mechanism of action.

Paracetamol (acetaminophen)

The analgesic paracetamol (acetaminophen) is the most investigated drug in terms of its mechanism of toxicity. The critical step is its conversion by several CYPs to a reactive imidoquinone (Figure 2.18), which is detoxified by reaction with GSH to form a paracetamol–GSH conjugate. Once GSH is depleted, the reactive metabolite binds to thiol groups on proteins (to form a 'protein adduct'), which may result in a loss of function of the protein. In combination with mitochondrial toxicity and effects mediated through cytokines and oxidative stress, this may contribute to the avalanche that leads to massive cell death at high doses of paracetamol.

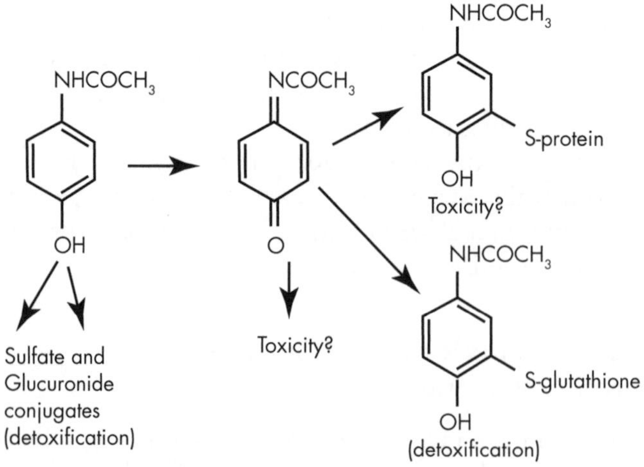

Figure 2.18 Paracetamol (acetaminophen) toxification and detoxification.

Tamoxifen

Tamoxifen is used as an anti-oestrogen to treat patients with breast cancer: it stops the growth of oestrogen-dependent tumours. However, it is activated to a DNA-reactive metabolite by conjugation and is a liver carcinogen in the rat, while its use has also been associated with an increase in endometrial cancer in women (although a causal relationship is not certain). In rats it is converted to a reactive sulfate conjugate, by a SULT related to SULT1A1 in humans, through sulfation of the α-hydroxy group. This leads to DNA adducts that are presumably responsible for the liver tumours in the rat (Figure 2.19).

Halothane

Halothane has been used for anaesthesia since diethyl ether became obsolete. It causes hepatitis with low incidence owing to the generation of a reactive intermediate by oxidative CYP metabolism. Halothane can also be metabolised by reductive CYP activity in which an electron is transferred to halothane (Figure 2.20). The immunogenic nature of the effect was indicated by rechallenging an affected individual

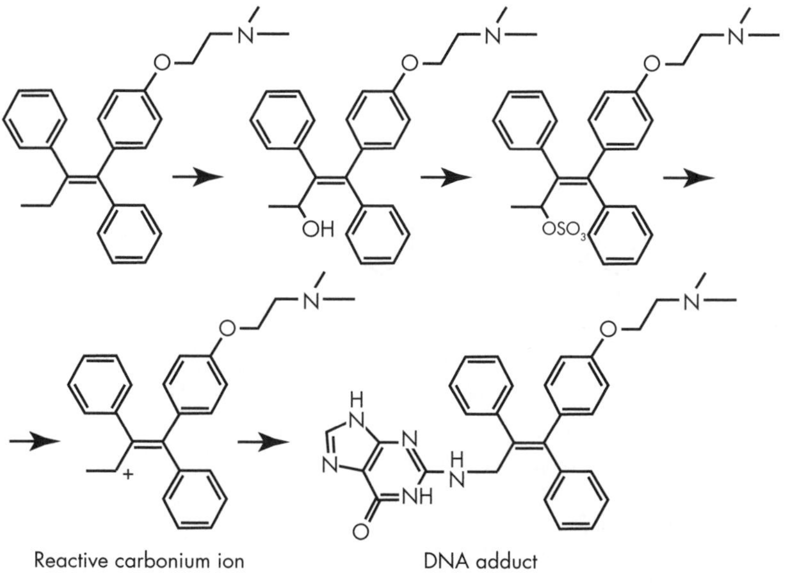

Figure 2.19 Bioactivation of tamoxifen by CYP and sulfation to a DNA adduct.

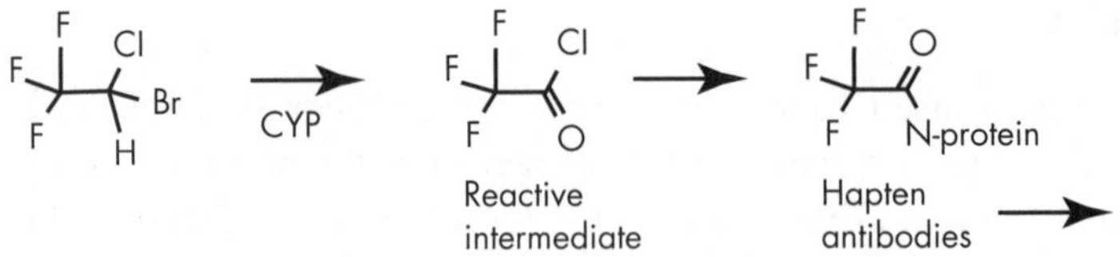

Figure 2.20 Toxification of halothane by CYP to a hapten.

with halothane: a more rapid and severe reaction was observed upon re-exposure. The presence of antibodies against oxidative halothane haptens in blood of sensitised patients confirmed that oxidative metabolism was responsible for the toxic reaction.

Acyl-glucuronides

Carboxylic acids can be converted to reactive acyl-glucuronides, as shown by the fact that the acceptor group may shift from the 1 position of glucuronic acid to the 2 or 3 positions. Subsequently, the ring may open to form a Schiff base, which can form protein adducts to lysine groups in proteins. These may be recognised by the immune system as haptens, resulting in an immune reaction. The reactive acyl glucuronide of diclofenac or zomepirac may react with GSH to give an S-acylglutathione conjugate that is still reactive; this is observed as a zomepirac metabolite in rat bile (Figure 2.21).

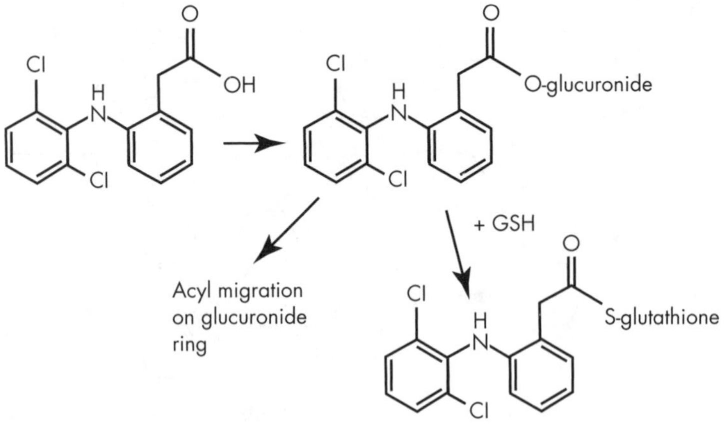

Figure 2.21 Formation of a reactive acyl-glucuronide from zomepirac.

Troglitazone

Troglitazone was introduced as a promising antidiabetic drug but had to be withdrawn from the market within a few years. It appears that its sulfate conjugate inhibits bile salt transport from the hepatocyte, leading to severe idiosyncratic hepatotoxicity.

Drug interactions: biotransformation

Drug interactions can occur at the pharmacodynamic (e.g. receptor) or the pharmacokinetic level. In particular, competition for biotransformation enzymes results in many potentially toxic interactions. If the metabolism of a drug is inhibited, its concentration may become too high, leading to typical overdose side-effects. On the other hand, if its metabolism is induced, the concentration drops below the required level and loss of therapeutic action may result in problems. During drug development, interaction studies may be required once the main metabolism of a medicine has been identified. However, even minor reactions may be toxicologically highly relevant if they lead to the formation of a toxic intermediate: if such a pathway is induced it may increase toxicity. Similarly, if a major pathway is inhibited, more of the drug may be channelled to a minor pathway, thus increasing toxicity associated with that route. In the product information leaflet, but especially in the Summary of Product Characteristics (SPC), extensive information is given about interactions.

Most attention has been given to CYP interactions because CYPs are the major Phase 1 metabolising enzymes, and because many interactions at that level have been identified. For instance, (S)-warfarin has to be administered at a carefully managed dosage because on one hand blood clotting should be prevented, while on the other hand clotting should be reduced to normal but not become severely impaired. (S)-Warfarin is metabolised by CYP2C9. If a patient on a steady-state dosing schedule starts therapy with a CYP2C9 inducer such as rifampicin, the dose of the warfarin must be adapted (i.e. increased) until the additional therapy stops, and the warfarin dose has to be reduced again. The opposite applies when an inhibitor of CYP2C9 is administered. Unfortunately, plant products may also contain CYP inhibitors: St John's Wort, which is widely used, shows strong interactions with CYP. In 1997 mibefradil was introduced as a calcium antagonist. It showed strong inhibitory action against CYP3A4 and CYP2D6, responsible for the metabolism of many drugs. In spite of the warnings in the

SPC, many fatal interactions with other drugs occurred, and it had to be withdrawn from the market within a year.

Much less attention is given to potentially toxic interactions for conjugating enzymes. It has been argued that drugs that are not metabolised at all but are excreted unchanged would be preferred, since they would not be subject to biotransformation interactions.

Further reading

Klaasen CD, ed. (2001). *Casarett & Doull's Toxicology. The Basic Science of Poisons*, 6th edn. New York: McGraw-Hill, chapter 6.

3

Molecular and cellular mechanisms of toxicity

J Fred Nagelkerke and Bob van de Water

Toxicity of xenobiotics can be studied at different levels: in the whole organism; in specific organs or cells from the organs as well as related cell lines *in vitro*; and finally at the molecular level. Toxicity of xenobiotics is in most cases due to an interaction with some cellular constituent(s) that ultimately affects certain cellular functions. This leads to either dysfunctioning or destruction of the cell. When looking for the *mechanism of toxicity* of a drug, therefore, the aim is to identify this critical first interaction that leads to an avalanche of follow-on effects, finally resulting in the toxic effect observed at the organ or whole organism level.

Much of the research on mechanisms is done on cell lines or primary cells in culture. In general, *in vitro* it is easier to identify and confirm a mechanism for a certain effect at the molecular level. Moreover, the cells can be genetically engineered to express certain enzymes (including conditional expression); similarly, a particular enzyme can be knocked out by siRNA for instance.

However, ascertaining the molecular mechanism of a drug-induced functional deficit requires confirmation *in vivo*, with an intact organ in its natural environment. In other words, such *in vitro* results have to be validated *in vivo* in order to establish their relevance. This is essential because many *in vitro* findings will be artefacts as isolated cells have lost their natural protective systems. To mimic the *in vivo* situation to some extent, co-cultures can be used in which cells under investigation grow together with a partner cell line, e.g. hepatocytes together with NIH3T3 fibroblasts.

This chapter focuses on events at the molecular and cellular levels. Some of them are very similar to pharmacological interactions, which is not surprising because pharmacology aims at the therapeutic application of interactions between a chemical and its receptor that are *in principle* toxic. Only in specific patients will this potentially toxic interaction work therapeutically, because in that person the effect may correct the disease process. That is also why safety requirements for drugs that are used *preventively* should be very strict, since in that case healthy people are being treated (for instance, oral contraceptive drugs).

This chapter is divided into three parts: molecular mechanisms of toxicity, cellular dysfunction and cell destruction.

Molecular mechanisms of toxicity

Virtually all biomolecules are potential targets for toxicants. These can be macromolecules such as proteins or nucleic acids, or small molecules such as lipids or the tripeptide glutathione. Interaction of toxicants with biomolecules can occur non-selectively, e.g. interaction with double bonds in polyunsaturated fatty acids or thiol groups of glutathione or proteins. In other cases the interaction is highly specific, e.g. with receptors on the plasma membrane or with promoter regions of DNA. Some compounds require biotransformation to become toxic (Chapter 2). Thus, the toxicant can react in different ways with its target molecule. Some major molecular mechanisms are outlined below.

Non-covalent, reversible binding to a receptor

Many toxicants act through binding to a receptor molecule, either a highly selective or a non-selective receptor. For instance, many effects in 'safety pharmacology' studies are due to binding of the toxicant to regular pharmacological receptors but result in unwanted side-effects. Early histamine H_1 antagonists were used against allergic reactions; they had unwanted sedative effects due to their effect on H_1 receptors in the central nervous system, while the therapeutic effect aimed at peripheral H_1 receptors involved in the allergic effects. Only when later, more polar H_1 antagonists were made that cannot pass the blood–brain barrier to the central nervous system, could the anti-allergic effect be obtained without the unwanted sedative side-effect. This sedative effect is due to interaction with the same receptor: a typical pharmacological side-effect.

If the steric conformation of the toxicant allows sufficient strong binding to a protein through hydrogen bonds or ionic bonds or hydrophobic interaction, it can activate or inactivate membrane receptors, intracellular receptors, ion channels or enzymes. The receptor theory of pharmacology applies to these unwanted side-effects. The cellular process that is initially affected can vary widely, from mito-chondrial energy production to proteins involved in signal transduction or regulation of the cell cycle.

Quite a few natural toxins act on such receptors and are or have been used therapeutically (e.g. curare, digitalis). Obviously, toxins may act through receptors that are not (yet) used therapeutically. In this case such receptors will not be known from pharmacology; nevertheless, their actions are similar to those of 'classical' receptors. The toxicological consequences are obviously mediated through these interactions with the receptor. A current problem is the very promiscuous binding of drugs to the so-called I_{Kr} channel, increasing the risk of QT prolongation and resulting in sudden heart death (torsade de pointes) (see Chapter 4).

Cationic drugs (in many cases amines) may bind to anionic sites, leading to toxicity; one example is binding to the pigment melanin. Chloroquine is such a drug, which by binding can accumulate massively in the pigmented areas (uvea) of the eye. This causes damage to the retina during long-term treatment of rheumatic diseases.

One reason why protein-type biologicals such as (humanised) anti-bodies apparently have relatively few side-effects may be that their bulky structure allows few strong interactions with other proteins in the body (exterior to cells). They may of course be (very) antigenic, which is currently a major concern with these biologicals.

A recent development is the binding of small siRNA fragments to RNA in the cell. This leads to the (irreversible) break down of the targeted RNA. This development promises the application of novel small RNA-type drugs that lead to breakdown of specific mRNAs and thus to knockout of specific gene products.

Covalent binding: irreversible interactions with receptors

Other toxicants bind irreversibly to their targets. These may be macro-molecules or low-molecular-weight targets. In most cases the drug has to be converted to a reactive intermediate by bioactivation (see Chapter 2) before it can bind covalently.

Binding to DNA Many drugs that cause cancer do so by covalent interactions with nucleophilic groups on DNA bases. The DNA adducts thus generated may cause mutations leading to cancer if they occur at certain critical positions (Chapter 5). In other positions the DNA adducts or DNA cross-links will lead to cell death. This is a therapeutic effect if the cell concerned is a cancer cell: an example of toxicotherapy!

Binding to proteins Reactive electrophilic metabolites of drugs may bind non-specifically to thiol or amine groups of proteins. As a result, the function of the protein (an enzyme, carrier protein, structural protein, etc.) may be lost: if sufficient molecules are thus incapacitated, the cell may lose control and die by apoptosis or necrosis. When the cell can be replaced by regeneration (e.g. in the liver), this may not be detrimental. However, loss of cells that cannot easily be replaced, as in the central nervous system, may cause loss of function of the organism. Methanol may destroy vision through toxicity in the retina and optic nerve.

Covalent binding to a protein may also lead to the formation of a *hapten*, resulting in immunologically mediated cytotoxicity if that protein is present on the cell surface. This is the case with halothane (see Chapters 7 and 10).

Binding to low-molecular-weight compounds Binding to a low-molecular-weight component such as the thiol group of glutathione may lead to exhaustion of that compound. Since glutathione has important protective functions, this is detrimental to the cell (e.g. paracetamol toxicity, Chapter 7).

Incorporation into endogenous compounds

A toxicant can be mistakenly incorporated into endogenous metabolite pathways, leading to disturbances. This is used therapeutically for certain anticancer or antiviral drugs by so-called 'anti-metabolites'. Another example is the 'lethal synthesis' of fluorocitrate upon administration of fluoroacetate. Because fluoroacetate is so similar to acetate, it is built into fluorocitrate, which subsequently blocks the citric acid cycle. This was discovered because some plants contain fluoroacetate, which killed cattle that fed on them.

Cellular dysfunction

Once a chemical has changed some molecular process inside the cell, how the cell or the organism responds depends on the nature of this change. If the change is decreased synthesis of a certain essential export protein, the cell itself may show little damage. In that case the toxicity may first be expressed elsewhere, by a lack of that particular protein. Similarly, a highly specialised function of the cell may be affected selectively, such as the production of a hormone or an essential metabolite.

These examples stress that *in vitro* assays with single cell types may miss relevant toxicity if a function is affected that has no direct consequences for the cell itself.

Another example of a change in cell function not leading to cell death but potentially having profound consequences for the organism is modification of signal transmission in nerve or neuronal cells. Such effects can lead to altered conduction of electrical impulses due, for example, to direct effects of toxic compounds via receptors or ion channels. This example shows the importance of measuring intact physiological systems for their loss of function due to selective toxicity at one particular site. It makes experiments in whole animals unavoidable as well as, ultimately, observations in patients in order to assess the significance of animal toxicity.

In many cases, though, the primary effect will have consequences for cell survival. This is obvious, for instance, for those compounds that affect mitochondrial energy production or membrane carriers involved in ion homeostasis. Consequences then are that certain cellular ion concentrations, e.g. that of calcium, become off balance, that the ATP content decreases below a critical level, or that processes that protect against oxidative damage become deficient.

Impaired cellular homeostasis

Loss of cellular homeostasis is the most severe final condition that ultimately leads to cell death. Cells contain a dedicated set of enzymes and processes that regulate various kinds of intracellular parameters such as ion concentration, ATP/ADP ratio, redox potential, mitochondrial and plasma membrane potential and DNA integrity. When toxicants bind to any of the proteins that regulate one of these control systems, this will affect homeostasis.

Oxidative stress and lipid peroxidation

Toxicants can impose oxidative stress on cells by either a direct or an indirect effect. Some drugs that are metabolised by cytochrome P450 lead to redox cycling. This is the case for paraquat and bleomycin, with which superoxide anion is generated continuously. Reactive oxygen species (ROS) are a collection of reactive forms of oxygen, some of which are radicals. When an oxygen molecule picks up one electron, the highly reactive superoxide anion is formed (reaction 1). This is a radical that may initiate lipid peroxidation. Since its production will also occur under normal physiological conditions, the body has excellent defence systems to remove the superoxide ion as fast as possible. The responsible enzyme is superoxide dismutase, which is one of the enzymes with the highest catalytic efficiency (reaction 2). However, in the presence of several metal ions, for instance ferrous ions, the Fenton reaction can occur, which converts superoxide to the even more reactive hydroxyl radical (reaction 3). Another form of reactive oxygen is the singlet oxygen atom.

$$O_2 + e^- \rightarrow O_2^{\cdot-} \text{ (superoxide anion)} \tag{1}$$

$$2O_2^{\cdot-} + 2H^+ \rightarrow O_2 + H_2O_2 \tag{2}$$

$$H_2O_2 + Fe^{2+} \rightarrow HO\cdot + OH^- + Fe^{3+} \tag{3}$$

These ROS are the cause of several forms of toxicity as a result of their radical character. At the same time they form a very important defence system during phagocytosis of bacteria, for instance. Under normal conditions, ROS are cleared from the cell by the action of superoxide dismutase (SOD), catalase, or glutathione (GSH) peroxidase. In this process glutathione, ascorbic acid and vitamin E (collectively called endogenous antioxidants) also play a role as an integrated detoxifying system. Under conditions of severe ROS production this protective system may fail because one of the components becomes rate limiting, leading to expression of toxicity. This complex scavenging system probably originates from the need to eliminate ROS that are produced during normal respiration coupled to oxidative phosphorylation. Because of the Pauli principle, the four electrons that are transferred from carbon to molecular oxygen during respiration are transferred one by one. Thus, during a short period superoxide molecules exist during normal respiration. It is estimated is that a few per cent of these molecules escape and are subsequently eliminated by the scavenging systems.

Major damage to cells results from the ROS-induced alteration of lipids such as polyunsaturated fatty acids in membranes, essential proteins, and DNA.

When lipids are the target, *lipid peroxidation* will occur. In this process an oxygen radical (it can also be a radical formed directly from a toxicant, as long as it can abstract a hydrogen atom from a lipid molecule) leads to an avalanche of a self-propagating radical reactions in which the fatty acids themselves turn into radicals (Figure 3.1). As a result, many peroxy fatty lipids are formed, which leads to the destruction of membrane integrity. The presence of sufficient vitamin E in the

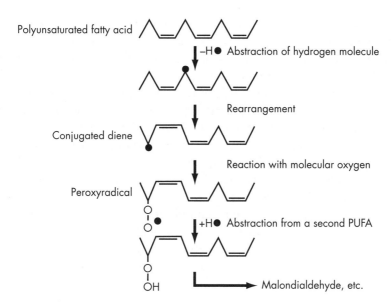

Figure 3.1 Lipid peroxidation. Reactive oxygen species or free radicals from toxicants abstract a hydrogen atom from polyunsaturated fatty acids (PUFA) in the cellular membrane. After rearrangement and reaction with molecular oxygen, a peroxy radical is formed. This in turn can abstract another hydrogen atom from a second PUFA; thus, the process is self-propagating. Ultimately the lipids are destroyed, to form at least 30 products. One of these is malondialdehyde, which is often used to determine the extent of lipid peroxidation in cells or tissues. Ethane is another product, which may be exhaled *in vivo*.

membrane can stop the propagation of the radical chain reaction, but vitamin E may become depleted during massive lipid peroxidation.

Lipid peroxidation will ultimately damage membranes beyond repair, resulting in loss of the organelle of which it is part or of the intact cell if it is a cellular membrane. Indicative of lipid peroxidation is the presence of lipid breakdown products (e.g. thiobarbiturate-positive material: TBARs) or (*in vivo*) ethane in exhaled air.

Sustained rise in intracellular calcium

The calcium concentration in blood is approximately 1.4 mM, while in the cytosol of cells the concentration is around 150 nM; thus, there is a 10 000-fold difference between the inside and the outside of the cell. The steep gradient allows calcium to act as a so-called second messenger in the cell. This gradient is essential for cell survival and cells therefore contain a number of systems to maintain this gradient. The plasma membrane contains several calcium pumps to rapidly extrude calcium from the cytoplasm; in addition, the endoplasmic reticulum and mitochondria function as storage sites for calcium. As the calcium concentration is crucial for the regulation of a large number of proteins, this concentration is under tight control (Figure 3.2). Interference of toxicants with this control system can lead to impairment of the cell's ability to maintain structural and functional integrity, depletion of ATP and microfilament dysfunction, leading ultimately to membrane blebbing and rupture. The fact that many proteins and enzymes are tightly regulated by free intracellular calcium contributes to these effects. For example the protein calmodulin is activated by calcium and thereby regulates the function and activity of other proteins and enzymes. Also, several enzymes with hydrolytic activity are regulated directly by calcium, including some proteases such as the calpains and nucleases.

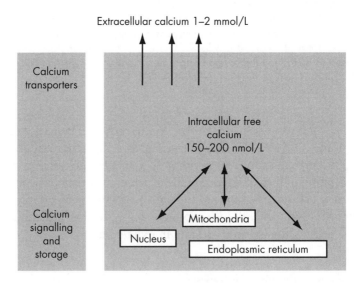

Figure 3.2 Intracellular calcium. The intracellular free calcium concentration is under tight control. Inside the cells the concentration is 10 000 times lower than outside. To maintain this difference, several pump systems in the plasma membrane, mitochondrial membrane and endoplasmic reticulum transport calcium out of the cell or into intracellular storage sites. When needed, calcium is pumped out again from the storage sites to act as a second messenger.

Impaired ATP synthesis

Loss of ATP results in cell death. Therefore, if inhibition of oxidative phosphorylation affects ATP synthesis excessively, this is lethal for cells. This can be due to inhibition of enzymes of the citric acid cycle, inhibition of components of the electron transport chain, all forms of hypoxia, and inhibition of ATP synthase function. Effects on the mitochondrial membrane potential through interaction with its components may also lead to cell death. Components leaking from mitochondria may initiate apoptotic cell death (see below). Several anti-AIDS compounds are such 'mitochondrial toxins'.

DNA damage

Binding of toxicants to DNA can lead to mutations and ultimately to uncontrolled proliferation of cells, i.e. cancer. This is discussed in Chapters 5 and 6. When the damage is too extensive, the repair capacity is insufficient. In this case the p53 system activates the cell destruction system called apoptosis, removing the cell from the organ. However, when the p53 system is itself mutated, the DNA damage sensing pathway and cell cycle arrest signalling do not work properly. This can result in hampered repair of DNA damage and proliferation of cells with an altered DNA profile. Indeed, a mutated form of p53 has been found in more than 60% of human cancers, indicating the very important role of this enzyme in repair of DNA-damage and prevention of cancer.

Cell destruction

Toxicants can induce two forms of cell death: necrosis and apoptosis. In necrosis, cells swell, rupture and release their contents into the body, a process accompanied by inflammation. Cells can also die through a cell suicide process called apoptosis. During apoptosis, the cell initiates a genetic programme leading to its own death. When cells die through necrosis, the release of intracellular compounds leads to an up-regulation of the immune system. But the billions of cells that undergo apoptosis every day do not induce an immune response: basically it is a physiological process.

Differences between necrosis and apoptosis

The main difference between necrosis and apoptosis is that necrosis is a *passive* process while apoptosis is an *active* process. In other words, necrosis occurs when the effect of the toxicant is so strong that the cell is irreversibly damaged and simply falls to pieces. The induction of apoptosis is much more subtle and allows the cell to execute its own death. Importantly, during apoptosis so-called apoptotic bodies are formed that contain the cellular components, (Figure 3.3). These bodies are engulfed by neighbouring cells or professional macrophages, which explains the lack of an immunological response. Examples of conditions that induce apoptosis and/or necrosis are shown in Table 3.1. The two

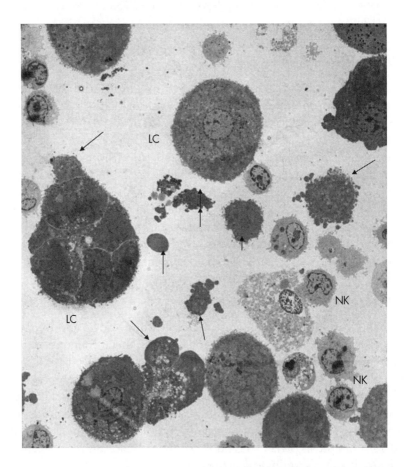

Figure 3.3 Apoptosis. After induction of apoptosis, cells do not simply fall apart but apoptotic bodies (arrows) are formed after 'blebbing' of the cell membrane. This micrograph shows the attack of natural killer cells (NK) on liver cells (LC). Several phases of apoptosis can be observed. The liver cell in the middle is still intact, while that on the left has already started to form apoptotic bodies. Other liver cells have completely disintegrated into small apoptotic bodies.

Table 3.1 Triggers of apoptosis and necrosis

Apoptosis	Necrosis
Inhibitors of kinases	Inhibitors of mitochondria
Tumor necrosis factor	Pro-oxidants
Calcium ionophores	Calcium ionophores
Ionising radiation	
Chemotherapeutics	
Chemical toxicants	
Withdrawal of growth factors	
Loss of cell adhesion	

most frequent conditions that induce necrosis are linked solely to (1) severe mitochondrial dysfunction leading to a collapse of the ATP production of the cell, or (2) increased formation of reactive oxygen species that ultimately leads to lipid peroxidation. Necrosis can easily be measured by analysis of enzymes (e.g. transaminases) in the blood.

Apoptosis is a highly regulated form of cell death distinguished by three phases: an initiation phase, a decision phase, and an execution phase. During the initiation phase the toxicants interact with their targets and the apoptotic machinery is activated. Then, during the decision phase, the balance between pro- and anti-apoptotic members of the Bcl-2 family determines whether the apoptotic process will proceed or be prevented. During the third step a family of cysteine-aspartate proteases (caspases) that cleave various proteins is activated, resulting in morphological and biochemical changes characteristic of this form of cell death. Simultaneously, DNA is cleaved by a family of endonucleases that cleave the DNA preferentially at the internucleosomal regions, leading to chromatin condensation; this is dependent on caspase activity. Microscopical examination of DNA-stained apoptotic cells reveals that DNA is present in tiny clumps at the border of the nuclei, often in the form of half-moons. Electrophoresis of fragmented DNA on an agarose gel results in a characteristic ladder-like pattern reflecting mono- and polynucleosomal fragments of $180 \times n$ base pairs.

Mechanism of induction and regulation of apoptosis

A large number of interactions of toxicants with cellular constituents can lead to initiation of apoptosis, e.g. binding to receptors, binding to ion channels, binding to DNA, binding to enzymes, binding to or oxidation of glutathione. However, the next step, the *decision* step, is quite similar for all toxicants. The Bcl-2 family members have the most important role in this phase. Bcl-2 and Bcl-xL are the two most important anti-apoptotic members of the Bcl-2 family of proteins. Bcl-2 contains four so-called Bcl-2 homology domains (BH1–BH4), which are absolutely required for its survival functions. At present, three groups of the Bcl-2 family proteins can be distinguished: (1) the anti-apoptotic proteins (most of which contain a C-terminal membrane anchor and the four BH domains), like Bcl-2 and Bcl-xL; (2) the pro-apoptotic members (which lack some of the four Bcl-2 homology (BH) domains; e.g. Bax, Bak); and (3) the BH3-only proteins (which contain only the third BH domain, and are all pro-apoptotic; e.g. Bad, Bik, Bid, Bim). The relative levels of pro- and anti-apoptotic proteins determine the susceptibility of cells to apoptosis. Several members of this protein family are capable of forming death-promoting or death-inhibiting homodimers and/or heterodimers. Many death signals converge through BH3-only proteins to the mitochondria.

Abundant evidence supports a role for mitochondria in regulating apoptosis at the induction phase as well as the decision phase. Specifically, it seems that a number of death stimuli target these organelles and stimulate the release of several proteins, including cytochrome c. Once released into the cytosol, cytochrome c binds to its adaptor molecule, Apaf-1, which oligomerises and then activates pro-caspase-9. Caspase-9 can signal downstream and activate pro-caspases 3 and 7 (Figure 3.4). At the decision level the release of cytochrome c can be influenced by different Bcl-2 family member proteins, including, but not limited to, Bax, Bid, Bcl-2 and Bcl-xL. Bax and Bid potentiate cytochrome c release, whereas Bcl-2 and Bcl-xL antagonise this event. Increasing evidence indicates a role for the pro- and anti-apoptotic family members in the regulation of the formation of pores in (mitochondrial) membranes.

During the last step, the *execution* phase, the various apoptotic stimuli converge even more. The ultimate step is cleavage of pro-caspase-3 into the active caspase-3. This caspase splices a large number of proteins, which leads to the biochemical and morphological characteristics of apoptosis. These proteins include structural proteins as well as regulatory proteins in the cell.

Apoptosis is not induced only by toxicants. It is also the mechanism by which the 'killer' cells from the immune system eradicate virally infected and tumorigenic cells. An inflammation can also be caused by the formation of haptens. Cytotoxic T-lymphocytes (CTLs) and natural killer (NK) cells are highly effective killers cells. They attack their targets either by secretion of granules containing perforins and granzymes or

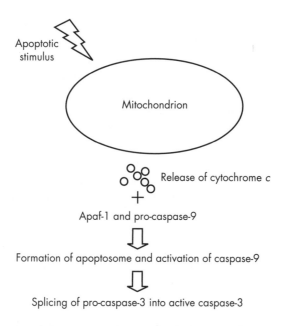

Figure 3.4 Mitochondrial damage and caspase-mediated apoptosis. Xenobiotics may affect the mitochondria, leading to release of cytochrome *c*; this will form a complex with apoptotic peptidase activating factor 1 (Apaf-1) and pro-capase-9. Ultimately this will lead to activation of caspase-3 and induction of apoptosis.

by binding to specific receptors present on the plasma membrane of the target cells. These receptors all belong to the TNF (tumour necrosis factor) superfamily and comprise the FAS, the TNF and the TRAIL (tumour necrosis factor related to apoptosis inducing ligand) receptors. The uptake of the granules ultimately leads to mitochondrial dysfunction and caspase activation. The regulation of the induction of apoptosis after binding to the receptors is more complex and includes trimerisation of the receptors, formation of an intracellular death domain, activation of caspase-8 and finally activation of caspase-3.

Determination of apoptosis

On the basis of the distinct biochemical reactions described above, several assays have been developed to detect apoptosis both qualitatively and quantitatively. The fragmentation of the nucleus can be determined easily with the microscope, the flow cytometer or by electrophoresis (PAGE). The activation of the caspases can be measured directly by determination of the formation of fluorescent fragments from specific substrates, by determination of splicing of the pro-caspases using western blotting, or by determination of low-molecular-weight fragments of caspase substrates such as PARP (poly(ADP-ribose) polymerase). The involvement of mitochondria can be determined using membrane-potential-sensitive dyes with time-lapse microscopy or flow cytometry. Also, during apoptosis phosphatidylserines present in the plasma membrane of cells re-orient from an inward orientation to a random orientation. The outward-oriented portion is then available for staining with Annexin V. Fluorescently labelled annexin V can then be used to determine apoptosis either with the microscope or with the flow cytometer. Measurement of apoptosis *in vivo* in normal humans or test animals is difficult, but the effects of compounds on blood cells can be determined easily using the Annexin V staining procedure. Biopsy is necessary for assessment of apoptosis in solid tissues. The tissue samples can be stained for DNA strand breaks caused by endonuclease activity using the so-called TUNEL assay.

Cellular responses to injury: stress kinases and molecular chaperones

When cells are injured by any of the mechanisms mentioned above, the damage is 'recognised' by signal transduction proteins. These include stress kinases such as the c-Jun N-terminal kinase (JNK). JNK is a kinase that in turn phosphorylates the transcription factor c-Jun. This results in the activation of c-Jun and the initiation of the transcription of a variety of genes. Genes that are expressed include those for proteins that are involved in the induction of apoptosis. JNK is not the only protein kinase that is activated; other protein kinases will activate other transcription factors, resulting in the expression of other genes. There are approximately 500 different human protein kinases. Depending on the level of injury, a particular pattern of protein kinase activation will result in a pattern of gene expression. For example, when mouse embryonic stem cells are exposed to DNA-damaging drugs such as cisplatin, etoposide or doxorubicin, the expression of around 3000 different genes is changed. Interestingly many of these genes are involved in the regulation of apoptosis, while other genes are related to DNA repair, cell cycle arrest or cytoskeletal organisation.

Toxicants may also affect the proteins in such a way that they become malfolded. This protein perturbation is recognised by molecular chaperones such as the family of heat shock proteins (HSPs). In normal cells, HSP70 is bound to the transcription factor heat shock factor (HSF). When proteins are damaged, HSP70 binds these malfolded proteins and tries to repair them. HSF is released at the same time; this is targeted to the nucleus, where it induces the transcription of heat shock protein family members that enable further repair of the cells.

Switch points between cell survival and apoptosis or necrosis

Cells will change the expression of proteins for two main reasons: to repair injury for survival or to promote the onset of apoptosis. This is a subtly balanced path; when repair is faster than the triggers for the onset of cell death, cells will ultimately stay alive. However, when cells try to repair the injury but in the long run are incapable of recovering completely, they will have a continuous stress response signalling that promotes apoptosis. In this case the induction of apoptosis will eventually win.

When cells are exposed to toxicants, the injury to a subpopulation of the cells can be so severe that they will die by necrosis. With increase in the concentration, the type of cell death will further switch from apoptosis to necrosis for more of the cells. This also directly indicates that, for a typical exposure to toxicants, apoptosis and necrosis can be seen at the same time. The switch from apoptosis to necrosis is controlled on the one hand by the levels of ATP. When ATP drops below around 10% of the normal values, cells will die by necrosis. On the other hand, when oxidative stress is so high that the antioxidant systems in the cells cannot deal with it, lipids will be peroxidised, followed by membrane leakiness, a hallmark of necrosis. Inhibition of necrosis by antioxidants can switch cell death from necrosis to apoptosis. If we block apoptosis by inhibition of caspase activation, do we then also fully protect against cell death? In general this is not the case for exposure to toxicants, since inhibition of caspases inhibits the induction of apoptosis but will eventually still lead to the onset of necrosis.

Further reading

Cory S, Adams JM (2002). The Bcl-2 family: regulators of the cellular life-or-death switch. *Nat Rev Cancer* 2: 647–656.

Klaasen CD, ed. (2001). *Casarett & Doull's Toxicology. The Basic Science of Poisons*, 6th edn. New York: McGraw-Hill, chapter 3.

Riedl SJ, Shi Y (2004). Molecular mechanisms of caspase regulation during apoptosis. *Nat Rev Mol Cell Biol* 5: 897–907.

4

Teratology

Lennart Dencker and Bengt R Danielsson

This chapter deals mostly with teratology or developmental toxicology, a part of what is often termed *reproductive toxicology*. It includes effects on male and female reproduction as well as functional defects induced in the developing individual after the period of sensitivity to strictly teratogenic action (Figure 4.1). Most attention in this latter respect has been paid to neurodevelopmental disorders that are induced by many chemicals as observed in experimental settings but are also often part of developmental syndromes in human offspring induced by ethanol, tobacco smoking and the use of other recreational drugs, antiepileptics and retinoids, for example.

Reproductive toxicology in adults

Male reproduction

There is a consensus that fertility in males is decreasing in the Western world, and much research has been done on its relation to chemical exposure in adult men. There are regional variations, however, that are not easily explained by differences in chemical exposure. For instance, a gradient is observed from lower sperm counts in males in Denmark and Norway to higher counts going east to Finland and the Baltics. As a result, the concept of *testicular dysgenesis syndrome* has been introduced, including, in addition to low sperm counts, hypospadias, testicular retention in the abdomen and an increase in testicular cancer. There is now evidence that this syndrome, or components of it, has its origin in fetal life. Lifestyle factors (e.g. those of the mother during pregnancy), seem to be involved rather than environmental exposure to toxicants. What role (recreational) drugs play is not clear. This example may also illustrate first that changes in disease patterns (in this case testicular cancer) may be due to factors other than the chemical environment, and secondly that research may have focused

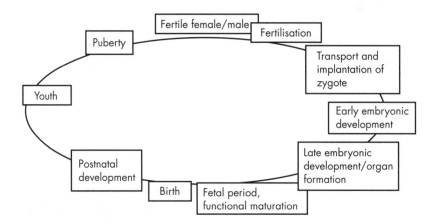

Figure 4.1 The reproductive cycle.

on factors at the wrong age – the adult testis instead of the developing testis.

Cellular targets of *toxicity in the adult male* are:

1. The *Leydig cell* (outside the blood–testes barrier), producing androgens under the influence of luteinizing hormone (LH). Leydig cell hyperplasia/neoplasia has been reported to be induced by a range of drugs during chronic treatment. Examples are oestrogens and antiandrogens, calcium-channel blockers, hypolipidaemic drugs (PPAR (peroxisome proliferator-activated receptor) ligands), histamine receptor blockers and dopamine agonists as well as antagonists, antihypertensive drugs, anticonvulsants, antiviral and antiprotozoal drugs, and cytostatics.
2. The *Sertoli cell* (part of the blood–testes barrier), a nurse cell for the spermatogenetic epithelium. Environmental toxicants such as dibromochloropropane (DBCP), dibromochlorotoluene and phthalates have been incriminated as Sertoli cell toxicants.
3. The *spermatogenetic epithelium*. It is reasonable to assume that this is the main target of cytostatic drugs and other chemicals that have effects on dividing cells. It should be noted that the spermatogenetic epithelium is in close contact with the Sertoli cells, making difficult exact identification of the target of many drugs. Cell surface FAS receptors and ligands on the two cell types are examples of interactions that regulate apoptosis of the spermatogenetic epithelium, which can be up-regulated by certain male reproductive toxicants.

The Sertoli cell has long been considered the target of phthalates, presently the most-studied chemicals regarding male fertility. Phthalates are not drugs, but were once important in medical technology because they were widely applied in plastic blood bags. In the past this led to high concentrations in cases of blood replacement in newborn babies, and newborns who undergo intensive therapeutic medical interventions are still exposed to higher concentrations than the general population. Recent research has prompted the notion of 'phthalate syndrome', characterised by alterations in androgen-dependent development of the entire male sexual tract, challenging the concept of the Sertoli cell as the only target of phthalates. In addition, other halogenated environmental chemicals such as polychlorinated biphenyls (PCBs), DDT (clofenotane, dichlorodiphenyltrichloroethane) metabolites, bisphenol A and alkyl phenols have been shown to affect development of the reproductive tract, often apparently at lower doses than those affecting sexual function in adults, again focusing on the role of environmental factors in fetal life.

Female reproduction

Several steps in ovarian follicular maturation, production and function of follicle-stimulating hormone (FSH), LH and steroids at various phases of the oestrous cycle may be targets of reproductive toxicants. The processes of ovulation, fertilisation of the egg in the oviduct, transport of the zygote in the oviduct, and finally implantation in the prepared endometrium may be affected. Smoking (notably exposure to poly-aromatic hydrocarbons) can cause precocious menopause, very likely due to aryl hydrocarbon (Ah) receptor mediated apoptosis of oocytes, in addition to long-term induction of various CYPs in the ovary. Nonsteroidal anti-inflammatory drugs (NSAIDs) appear to inhibit prostaglandin-dependent processes such as ovulation and implantation, causing reversible infertility in women on such treatment. In addition, as for males, cytostatic drugs are strongly toxic to female reproduction. Finally, although the effect of centrally acting drugs (e.g. antidepressants and ethanol) on libido is most studied in males, suppression of libido most likely occurs in females too.

Teratogenesis

Today there are some 20 drugs on the market that cause structural malformations in offspring if they are used during pregnancy. Many more have a presumably more or less reversible pharmacological action in the fetus similar to that in the adult. For some of those causing morphological malformations, the pertinent mechanisms of action are fairly well known, and often these correspond to the mechanisms of the wanted or expected pharmacological actions or secondary pharmacological actions in the mother. The difference is that the downstream effects in the embryo or fetus may lead to detrimental errors in morphological development and physiology. This understanding provides hope in avoiding the development of teratogenic drugs, but at the same time it is discouraging because we cannot prevent the passage to the embryo of a drug that is given to the mother to exert a pharmacological effect but that is dangerous to the embryo.

Since most drugs have specific and individual modes of action as teratogens, mechanisms are discussed below in connection with each teratogen or group of teratogens.

In contrast to the knowledge of mechanisms today, in the late 1950s and early 1960s when the extremely teratogenic drug thalidomide was put on the market, little was known or even expected as regards threats to the embryo or fetus. Only such strong insults as irradiation from X-rays and radioactive sources, specific infections, a few metabolic diseases, and cytostatic and hormone treatment of the mother were at that time known to cause severe malformations.

Understanding of the mechanisms of action of teratogens is not a purely academic scientific matter; it is essential for the development of future drugs without teratogenic properties. This is of major importance in some diseases, such as epilepsy, where the widely used antiepileptic drugs today are established human teratogens.

Structure–activity relationships and so-called *class effects* apply in teratology as in other pharmacological and toxicological situations. There may be many more common pathways to aberrant morphological development in the embryo or fetus than we know of today. Examples are drugs that primarily or secondarily affect the haemodynamic physiology of the fetus, such as antiepileptic drugs. To some extent already, but hopefully much more in the future, the tools of molecular biology and bioinformatics will contribute to our understanding of (class-specific) mechanisms of teratogenic action and common pathways to aberrant development. Up to the 1980s, developmental biology was mostly a descriptive science. When techniques to describe spatial and temporal gene expression patterns in the embryo became available and were extensively applied to older concepts such as induction, tissue interactions, cell migration, gradients of morphogens and apoptosis, the results of this cross-fertilisation in understanding basic developmental biology became impressive and productive. In teratological sciences, this emerging knowledge is still only in its cradle.

In this chapter the focus is on drugs. This is the area where mechanisms are becoming increasingly more established, and this knowledge can eventually be applied to chemicals in general. The section Factors to Consider for Human Relevance in Teratogenesis below also considers conditions for other chemicals.

It must be pointed out that for all malformations occurring in the Western world, genetic factors are quantitatively more important than environmental factors. In a worldwide perspective, nutritional deficiencies are also of great importance, such as those of iodine (cretinism and mental retardation), folic acid, zinc and vitamin A. This has relevance for chemically induced disturbances of development as well, since chemicals can interfere in the metabolism of nutrients. Perhaps the best-known example in this setting are antiepileptic drugs, which can cause folate deficiency.

General principles of teratology

1. Whether the exposure leads to teratogenic effects depends on the developmental stage at the time of exposure; the effects are related to the maternal serum concentration of the drug.
2. True teratogens most often act on the conceptus, but the importance of a toxic effect on the mother or the placenta cannot be ignored.
3. Teratogens most often act in a specific way on developing cells and tissues to initiate sequences of abnormal developmental events. However, more recent research stresses the importance of changes in physiological and homeostatic conditions, such as haemodynamics and oxygen regulation of the fetus.
4. Susceptibility to teratogenesis depends on the genotype (polymorphism with respect to receptor levels, cytochrome P450s, etc.) of the conceptus and the manner in which genes interact with adverse environmental factors.
5. The nature of the drug, such as its physicochemical properties, determines whether a potentially teratogenic drug will get access to and influence the conceptus.
6. Four principal manifestations of deviant development can be distinguished: death, malformation, growth retardation and functional deficit.

7. Dose (concentration) dependency starts from no effect, and via growth retardation and malformations reaches concentration levels lethal to the embryo. This is important, because too low or too high (embryolethal) doses may hide the true teratogenic potential of a drug.

Animal models

Hundreds of chemicals have been shown to produce developmental toxicity in experimental animals; yet only a handful of compounds, notably pharmaceutical and recreational drugs, are known to cause malformations in human embryos. This discrepancy in numbers is probably not a matter of real species differences but of differences in exposure: experimental animals are often exposed to concentrations far above those occurring normally in the environment or in drug-treated patients. This is not to say that such experiments are worthless: they may give important mechanistic information, and they contribute to the hazard identification and risk assessment procedure for drugs and other chemicals. In the clinical situation, some individuals may also be exposed internally to very high levels as a result of genetic differences in metabolism, diseases or interactions, for example simultaneous exposure to another compound that has a similar mechanism of action, or the same enzyme for its biodegradation.

Are there reliable and predictive animal models for all compounds known to cause malformations in human offspring? The answer is yes, although the models are not necessarily quite straightforward. For instance, for thalidomide, monkeys have to be used; and for angiotensin-converting enzyme (ACE) inhibitors and warfarin, rodents have to be exposed postnatally rather than prenatally. With valproic acid, neural tube defects in humans occur *posteriorly* rather than *anteriorly* (spina bifida). In fact, the relevant animal model has often been developed only after the drug was discovered to be a human teratogen. On the other hand, these animal models are useful for our understanding of mechanisms, and may prove predictive for future drugs.

Factors to consider for human relevance in teratogenesis

Sometimes chemicals have to be classified with respect to developmental toxicity in the absence of relevant data and/or the data normally required for registration of drugs or pesticides. One then has to take an overall perspective, considering data from experimental animals as well as humans, as outlined below.

- Are there case reports or (even weak) epidemiological studies?
- Is a structure–activity relationship likely for known teratogens?
- Have the route of administration, the number and severity of disturbances, and how disturbances are distributed among litters been investigated reliably?
- Is the toxicokinetics in test animals similar to that in humans?
- Is the mechanism of action, if known, relevant to humans?
- Are there disturbances also in the control group?
- Maternal toxicity is often blamed, but is it certain?
- Are disturbances in treated and/or control animals that may be just 'variants' (e.g. in skeletal development) rightly or wrongly blamed on maternal toxicity?
- Is there a reduction in fetal weight?
- Are malformations severe/rare? They may be due to genetic variation, but should still be considered.
- Is the dam affected during gestation or lactation?
- Are there differences in postnatal development? If so, can the dam nurse the pups (i.e. give milk, take care)?

Pregnancy labelling for drugs

All drugs in the European Union are assessed concerning risks when used in pregnancy. The labelling is based on an integrated assessment of available human pregnancy outcome data in addition to results in animal reprotoxicity and teratology studies. Effects are listed from the most severe effects (1) down to no effects (8).

1. The drug causes/is suspected to cause serious birth defects when administered during pregnancy.
2. The drug has harmful pharmacological effects on pregnancy and/or the fetus/newborn child.
3. There are no adequate data from the use of the drug in pregnant women. Studies in animals have shown reproductive toxicity. The potential risk to humans is unknown. **Or:** Animal studies are insufficient with respect to effects on pregnancy and/or embryonic/fetal development and/or parturition and/or postnatal development. The potential risk for humans is unknown.
4. No clinical data on exposed pregnancies are available for the drugs. Animal studies do not indicate direct or indirect harmful

effects with respect to pregnancy, embryonic/fetal development, parturition or postnatal development.

5. Data on a limited number (n = ?) of exposed pregnancies indicate no adverse effects of the drug on pregnancy or on the health of the fetus/newborn child. To date, no other relevant epidemiological data are available. Animal studies have shown reproductive toxicity (specify). The potential risk for humans is unknown. **Or:** Animal studies are insufficient with respect to effects on pregnancy and/or embryonic/fetal development and/or parturition and/or postnatal development. The potential risk for humans is unknown.

6. Data on a limited number (n = ?) of exposed pregnancies indicate no adverse effects of the drug on pregnancy or on the health of the fetus/newborn child. To date, no other relevant epidemiological data are available. Animal studies do not indicate direct or indirect harmful effects with respect to pregnancy, embryonic/fetal development, parturition or postnatal development.

7. Data on a large number (n = ?) of exposed pregnancies indicate no adverse effects of the drug on pregnancy or on the health of the fetus/newborn child. To date, no other relevant epidemiological data are available.

8. Well-conducted epidemiological studies indicate no adverse effects of the drug on pregnancy or on the health of the fetus/newborn child. (In other words: the drug can safely be used during pregnancy.)

For groups 6–8, it must be recognised that a large number of pregnant women have to be studied in order to produce enough statistical power to exclude the risk of teratogenesis of a drug (see Chapter 1).

Some examples of drug-induced teratogenicity

Retinoids

Retinoids (vitamin A-like compounds) are used as medication for skin disorders such as severe cystic acne (isotretinoin) and psoriasis (etretinate), because they have a profound effect on cell differentiation in the skin. A simple view is that this is also why retinoids cause embryonic malformations. The retinoids are also used in therapy of certain forms of cancer. Retinoids act via a family of retinoic acid receptors (RARs) in the target organs. The dramatic effect of excess (as well as of deficiency) of retinoids on embryonic development is largely due to the fact that these same receptors are expressed in embryonic tissues and have a master role in controlling the expression in particular of regulatory genes in the embryo. Some of these genes (or gene products) in their turn are important for regulating genes important for laying down the body plan.

The syndrome of retinoid-induced malformations (Figure 4.2) includes brain damage, especially of the hindbrain area (Dandy–Walker

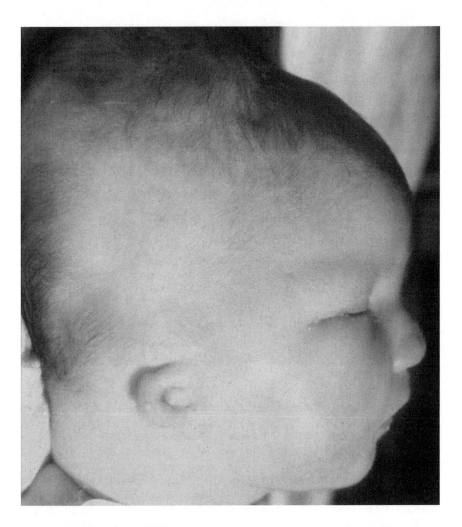

Figure 4.2 A newborn child exposed *in utero* to retinoic acid derivative (Accutane). Microtia is seen (right side), including absence of ear canal. The child also has a Dandy–Walker malformation of the brain (malformations especially in the cerebellum and fourth ventricle), transposition of the great vessels, and thymic hypoplasia. Published with the permission of Dr. Ed Lammer.

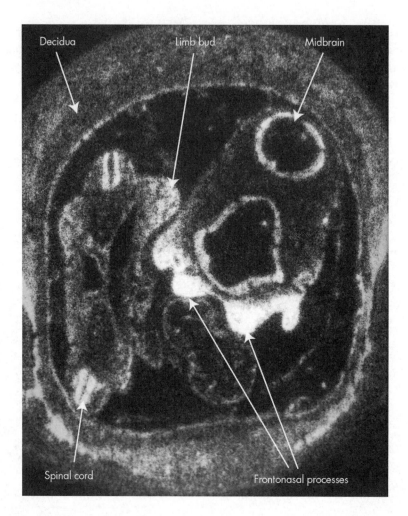

Figure 4.3 An autoradiogram showing the distribution of radiolabelled retinoic acid in a mouse embryo at the stage of great sensitivity to the teratogenic action of retinoic acid. The labelled retinoic acid had been injected to the mother four hours earlier, and thus passed to the embryo. In contrast to the gradient from caudally to rostrally along the body axis of endogenous retinoic acid discussed in the text, the administered retinoic acid accumulated in areas of the embryo expressing high levels of cellular retinoic acid binding protein (CRABP I). These areas include the frontonasal processes and branchial arches (not shown), which correspond to the severe facial and ear malformations shown in Figure 4.2, and of the brain, especially the hindbrain (not shown here). Reproduced from Dencker L, Annerwall E, Busch C, Eriksson U. (1990) Localisation of specific retinoid-binding sites and expression of cellular retinoic-acid-binding protein (CRABP I) in the early mouse embryo. *Development* 110(2): 343–352, with the permission of *Development*.

malformations), and aberrations in neural crest (NC) cell derivatives resulting in a small and asymmetric face, especially brachygnathia, small, malformed or absent external ear, atresia of the auditory canals, thymus hypoplasia, and conotruncal heart malformation (deficient separation of aorta and arteria pulmonalis).

Evidence that the normal functions of retinoids (notably retinoic acid) are disturbed, and not merely that retinoids exert general toxicity, e.g. by being chemically reactive, comes from a series of reports over a long period, showing that:

1. Vitamin A deficiency causes malformations.
2. Severe malformations occur in knockout mice lacking retinaldehyde dehydrogenase (Raldh2), in which the embryos cannot transform retinal (a retinol metabolite) into retinoic acid.
3. Thus, lack of the active ligand, as of (1) and (2), of the retinoic acid receptors causes malformations.
4. Malformations occur if a few or several members of the retinoic acid receptor family are genetically knocked out in different combinations.
5. Excess of active ligand(s) of the retinoic acid receptors, as in retinoid treatment, causes malformations.

The spatial expression of the rate-limiting Raldh2 for retinoic acid synthesis in the hind part of the embryo (high retinoic acid concentration), and of the retinoic acid-inactivating CYP26 in the anterior part of the embryo (low concentration), indicates that a gradient of retinoic acids is formed along the embryo axis under physiological conditions. Precise local concentrations, especially within the area of the hindbrain, may thus govern pattern formation. The syndrome of malformations involves hindbrain derivatives (e.g. the cerebellum) and structures dependent on NC-cell formation, migration and further development. The fact that these affected NC cells emanate primarily from the hindbrain makes this theory of the importance of local concentrations of retinoic acid attractive. It is very likely that exposure of the embryo to retinoids from the mother during retinoid therapy results in a far higher concentration than the physiological level in the embryo, and that this eradicates the embryo's own gradient(s). An additional finding is that cellular retinoic acid-binding proteins (CRABP I) with a high capacity for binding retinoids are specifically expressed in the hindbrain and in NC-derived cells (e.g. frontonasal processes, Figure 4.3). It remains to be shown what role, if any, CRABP I has in regulating the proposed gradient, in mediating inactivation of retinoids, or in shuttling retinoids to the receptors (RARs) in the nucleus.

Antiepileptic drugs

The antiepileptic drugs (AEDs) valproic acid, phenytoin, phenobarbital and carbamazepine as well as dimethadione (withdrawn from the market) are all established human teratogens. Despite the known teratogenicity of AEDs, women with epilepsy require treatment with AEDs during pregnancy (approximately 50 000 in Europe and North America alone every year), since the risk of seizure-related adverse effects (maternal death and adverse fetal effects) are considered to be more harmful than the twofold to threefold increased risk of drug-induced malformations.

Valproic acid (VPA) causes a different pattern of malformations both clinically and in animal studies from those of other antiepileptic drugs. VPA teratogenicity is mainly characterised by neural tube defects, notably spina bifida in human embryos, a smaller brain and changes in the facial expression, in addition to several less-frequent malformations. The different pattern of VPA teratogenicity suggests that its teratogenic mechanism is different from those of other AEDs. VPA was recently identified as a histone deacetylase inhibitor. When histones under the influence of VPA remain acetylated, they are less tightly associated with the DNA, making regulatory areas of genes more accessible to transcription factors and accessory, co-regulatory proteins, resulting in increased or decreased transcriptional activity, probably of hundreds of genes.

Since embryos of experimental animals, notably mice, exhibit neural tube defects upon exposure of dams to VPA, although these are anterior rather than posterior, realistic models for the teratogenic action of VPA are available. It has recently been shown that upon exposure to VPA a multitude of genes will be up-regulated or down-regulated that code for signal transduction, transcriptional and translational control, etc. However, genes are also affected that code for structural proteins and proteins engaged in cytoskeletal and extracellular matrix function and interactions, as well as a multitude of genes known to be important for neural tube closure, throwing light on the aberrant morphological development of the embryo. It remains to be shown which genes are the most fundamental in the mechanism(s) of teratogenesis; they may well be some of the genes under the influence of histone deacetylase.

Phenytoin (PHT), *phenobarbital* (PB), *carbamazepine* (CBZ) and *dimethadione* (DMD) induce similar patterns of fetal adverse effects in both humans and animal models. The manifestations include growth retardation, minor defects related to growth retardation (e.g. micro-cephaly and hypoplasia of certain skull bones) and structural malforma-tions, such as orofacial clefts, cardiovascular malformations and distal digital reductions. In view of the similar pattern of defects, a common mechanism for the teratogenic action of these AEDs is likely. Much attention has been paid since the 1970s to the possible role of reactive metabolites (epoxides), generated via the cytochrome P450 (CYP) system. However, there are several arguments against this theory, for example that CBZ and DMD lack the capacity to form epoxides and that the immature embryo essentially lacks the capacity for metabolisa-tion via the relevant CYP forms.

Accumulating evidence suggests the teratogenicity to be directly related to a common pharmacological property, resulting in cardiac arrhythmia and hypoxia in the embryo (Figure 4.4). PHT, PB, DMD and to some extent CBZ (but not VPA) all bind to a specific potassium ion channel expressed by hERG (human ether a-go-go related gene). The binding results in inhibition of a potassium ion current, the rapid component of delayed rectifying potassium ion current (I_{Kr}), at clinically

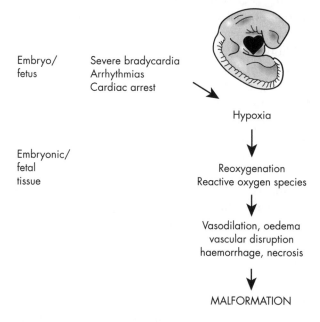

Embryo/ fetus — Severe bradycardia / Arrhythmias / Cardiac arrest

→ Hypoxia

Embryonic/ fetal tissue

↓

Reoxygenation
Reactive oxygen species

↓

Vasodilation, oedema
vascular disruption
haemorrhage, necrosis

↓

MALFORMATION

Figure 4.4 Mechanism for teratogenicity of phenytoin and other I_{Kr}-blocking drugs. Drug-mediated embryonic cardiac arrhythmia results in hypoxia/reoxygenation damage.

relevant concentrations. I_{Kr} plays a major role across species in repolarisation and cardiac rhythm regulation before the embryonic heart is innervated and the atria/ventricles are separated. This period corresponds approximately to weeks 4.5 to 10 of human pregnancy. When the primitive heart tube starts beating, the cardiac output and oxygen supply to the embryo are dependent on I_{Kr}-regulated peristalsis-like rhythmic contractions. In cultured rodent embryos, PHT, PB, DMD and CBZ (but not VPA) cause embryonic bradycardia and arrhythmia at clinically relevant concentrations. *In vivo* studies in rodents show that teratogenic doses produce severe embryonic hypoxia due to arrhythmia. Transient hypoxia caused by other means is teratogenic as well, and can induce all manifestations observed after *in utero* exposure to these AEDs. The concept of embryonic cardiac arrhythmia as a cause of malformations is compatible with the generation of reactive oxygen species following a period of interrupted oxygen supply, and it shows that reactive oxygen species play an important role in AED teratogenicity.

This mechanism provides a basis on which to explain various observed manifestations as follows.

- Long-lasting bradycardia results in prolonged embryonic hypoxia and correlates well with reported embryonic death, growth retardation, and minor structural abnormalities (e.g. decreased growth of skull bones).

- Episodes of embryonic arrhythmia result in episodes of hypoxia followed by generation of reactive oxygen species during the reoxygenation phase. Temporary hypoxia is a potent teratogenic factor that produces similar stage-specific orofacial clefts and digital reductions, preceded by similar early histological changes in embryonic tissues (oedema, vascular disruption, haemorrhage and necrosis).

- Arrhythmia results in alterations in embryonic blood flow and blood pressure. In experimental studies such alterations can produce the same type of cardiovascular defects as AEDs (septal defects and transposition and absence of vessels).

The risk of human malformations is greatly increased after concurrent treatment with two or more AEDs. Additive pharmacological effects (when both AEDs inhibit I_{Kr}; e.g. PHT and PB) and higher plasma concentrations due to displacement from plasma proteins (when the both AEDs are highly protein bound; e.g. PHT and VPA) are most likely important in explaining the highly increased risk after polytherapy with AEDs.

Potent hERG-blocking drugs

There is accumulating evidence that other drugs that inhibit I_{Kr} at clinically relevant concentrations are also teratogenic. The class III antiarrhythmic drugs dofetilide, ibutilide and almokalant were developed to inhibit I_{Kr} and produce similar stage-specific teratogenic effects in animal studies (cardiovascular defects, orofacial clefts, digit defects and, at higher doses, embryonic death). The malformations are preceded by embryonic bradycardia/arrhythmia due to hERG channel inhibition (Figure 4.5) and embryonic hypoxia (Figure 4.6).

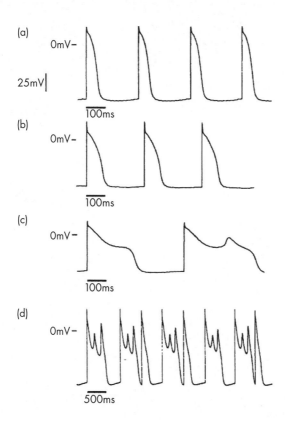

Figure 4.5 Transmembrane recordings of action potentials of the spontaneously beating embryonic rat heart. Increasing concentrations of almokalant result in brady-cardia due to prolongation of the action potential duration and the arrhythmia is associated with occurrence of early after-depolarisations. (a) Control; (b) 0.1 µmol/L almokalant; (c) 1 µmol/L almokalant; (d) 10 µmol/L almokalant. Reproduced from Abrahamsson C, Palmer M, Ljung B, *et al.* (1994) Induction of rhythm abnormalities in the fetal rat heart: a tentative mechanism for the embryotoxic effect of the class III antiarrhythmic agent almokalant. *Cardiovasc Res* 28(3): 337–344, with the permission of *Cardiovascular Research*.

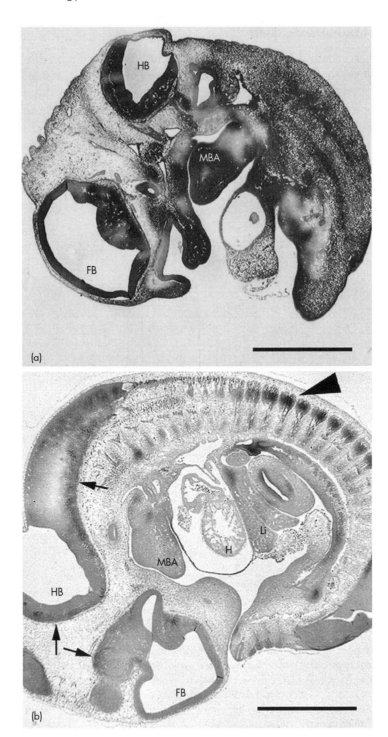

The macrolide antibiotics *erythromycin* and *clarithromycin* have recently also been shown to inhibit I_{Kr} as a side-effect at clinically relevant concentrations. In two large studies erythromycin has been associated with an increased risk of cardiovascular defects such as ventricular septal defects and vessel defects. In contrast, phenoxymethylpenicillin (penicillin V), which is used to treat the same type of infections as erythromycin in pregnancy but lacks I_{Kr}-blocking potential, does not increase the risk of malformations. The available human pregnancy outcome data for clarithromycin are too limited to allow any conclusions to be drawn on teratogenic potential. However, in animal teratology studies, clarithromycin produces the same pattern of developmental toxic manifestations.

ACE inhibitors and angiotensin (AT) antagonists

Most teratogens are detrimental primarily in the first trimester of human pregnancy. A few, however, can act only later when the physiological systems upon which they act have developed. Drugs that affect the renin–angiotensin–aldosterone system cause malformations, fetal morbidity and mortality. Thus, fetal malformations caused by ACE inhibitors and AT antagonists typically affect organs dependent on a normal blood pressure for their proper development, notably the kidneys and the calvarium, the membranous bone of the skull. In addition to this, it is not unlikely that local AT receptors have a role in development, independently of the haemodynamic system. Malformations have been observed after treatment in the 26th week, but most likely can occur earlier in gestation.

The syndrome of malformations and fetotoxicity caused by ACE inhibitors includes tubular dysgenesis in the kidneys, oligohydramnios

---

Figure 4.6 Images of rat embryos at day 14, immunostained with antibodies against a hypoxia marker (pimonidazole hydrochoride) previously given to the dams. Note immunostaining (darker areas) indicating severe hypoxia in the almokalant-treated embryo (a) as compared with control (b). Arrows in (b) indicate immunostaining, i.e. hypoxia, of the outer layer in the neural tube; the large arrowhead indicates immunostaining of the somites in controls. FB, forebrain; H, heart; HB, hindbrain; Li, liver; MBA, mandibular branchial arch. Bars represent 2 mm. Reproduced from Danielsson B, Skold AC, Johansson A, Dillner B, Blomgren B (2003) Teratogenicity by the *h*ERG potassium channel blocking drug almokalant: use of hypoxia marker gives evidence for a hypoxia-related mechanism mediated via embryonic arrhythmia. *Tox Appl Pharmacol* 193: 168–176, with the permission of *Toxicology and Applied Pharmacology*.

(decreased amount of amniotic fluid) and, most likely as a consequence of this, also hypoplasia of the lungs, intrauterine growth retardation, neonatal hypotension and oliguria/anuria. Hypocalvarium is a severe but rather rare additional deficiency, the brain being more or less exposed to the exterior. Various heart malformations and intrauterine closure of ductus arteriosus have been reported.

Warfarin

Exposure to warfarin during pregnancy, which seems to have occurred most often in women with artificial heart valves, entails a dual risk for the fetus. In the first place, severe haemorrhages may occur in the fetus, resulting in severe brain damage. Secondly, chondrodystrophy may occur, characterised by hypoplasia of the nose, brachygnathia, hypoplasia of the distal phalanges, stippling of uncalcified epiphyses, areas of disturbed cartilage formation resulting in premature calcification, spontaneous abortions, prematurity and stillbirth.

Similarly to the carboxylation of glutamic acid residues of prothrombin to form gamma-carboxylated prothrombin in the liver, in cartilage and bone so-called gla proteins mature by carboxylation. If this does not occur, disturbed development will result as above. The following is evidence that the pattern of malformations is due not to non-specific toxicity of warfarin but to its pharmacological effect. There exists a human mutation on the gene encoding vitamin K reductase. The phenotype of these children, sometimes termed pseudo-warfarin embryopathy, resembles that of warfarin-exposed children, exhibiting short nose and brachygnathia, distal digital hypoplasia and epiphyseal stippling on infant radiographs, all of which are virtually identical to features seen following first-trimester exposure to warfarin and other coumarin derivatives. In young rats also, warfarin exposure will result in short nose.

Diethylstilbestrol

The story of the effects of diethylstilbestrol (DES) in humans described below now seems very old and perhaps not relevant today. However, it illustrates the danger of using very potent drugs such as hormones during pregnancy.

In the late 1930s, DES began to be used as an oestrogenic compound in the USA and some European countries to treat women with impending miscarriage, the idea being that it would stimulate oestrogen and progesterone production in the placenta. Dosing started from around week 6 after conception, and the dose increased each week until the end of gestation, in total up to 16 grams being dosed during pregnancy. In 1971, the US Food and Drug Administration (FDA) issued a warning about the use of DES during pregnancy after a relationship had been reported between exposure to DES and the development of clear cell adenocarcinoma (CCA) of the vagina and cervix in the off-spring. This had been found in young women, in whom it is very rare as a spontaneous cancer. The cases were distributed in and around cities where the medical profession had recommended DES treatment during pregnancy. The number of new CCA cases per year increased for a number of years after 1972, but then stabilised and thereafter decreased. Since a few million fetuses were exposed *in utero* in the USA from the late 1930s to the early 1970s, one would perhaps not have expected such an early decrease in the number of new cases. However, since DES-induced CCA appears mostly from puberty up to the age of around 30 years, fewer new cases appear now, even though many women alive now were exposed *in utero*. The risk of getting CCA has been considered to be in the range of 1–10 per 10 000 prenatally exposed women.

Both female and male offspring of DES-exposed women exhibited aberrations in reproductive organs and functions at a far higher frequency than that of CCA. These include transversal vaginal septa and ectopically placed 'clear' cells in the vagina, most likely an aberration underlying the later CCA in a subfraction of the women. An alternative explanation of the CCAs may be a disturbed imprinting (methylation) pattern of genes in these hormone-dependent embryonic organs. Further malformations were cervical hypoplasia, small, T-shaped uteri, deformities and malfunctions of oviducts leading, among other effects, to extrauterine pregnancies, difficulty in becoming pregnant, and spontaneous abortions and preterm delivery, perhaps due to limited space in small uteri. It is evident that malformations in the reproductive tract also affect its physiology in adulthood.

Exposed male offspring developed cysts in the epididymis, small penis and testes, abdominal retention of testes and impaired semen quality and fertility as adults.

It is worth mentioning that animal experiments already performed when the drug was introduced to the market had shown DES to cause malformations in the reproductive tract of offspring when administered to pregnant rats. At that time, animal reprotoxicity experiments were most likely considered less predictive than we would consider them today.

Thalidomide

Thalidomide is currently coming back as a drug, now being used against leprosy and proposed for use against certain forms of leukaemia, as an immunomodulatory agent in lupus and AIDS, and possibly as an anti-tumour agent based on anti-angiogenic properties. Thalidomide was used as sedative drug and against nausea in pregnancy in the period from 1959 to 1962. It was then abandoned because it was shown to be teratogenic, around 10 000 children being born with severe malformations, including short or absent limbs (phocomelia or amelia). Had these rather characteristic and unusual malformations of the legs not occurred, even more children would have been born with the wide range of other abnormalities such as heart malformations and anomalies in intestines, kidney, ears and eyes that also occur. A child could be born without arms (sensitivity period around days 24–30 after conception) or legs (around days 28–33), or without both arms and legs, depending on the timing and duration of consumption of thalidomide. Thus, very early developmental events occurring in the limbs are affected. The lack of a proper animal model for this malformation makes a mechanistic approach difficult, and through the years some 30–40 mechanisms have been proposed. It is only recently when immunomodulation by thalidomide has been studied that new mechanisms have been considered, such as effects on signalling systems that may be shared between the immune system and embryonic structures.

It has been frustrating for the scientific community not to understand how a drug like thalidomide causes malformations. This has bearings on risk assessment, since there is uncertainty about the risks of similar new malformations in human embryos, and this is especially challenging because of the lack of good animal models for thalidomide, except for monkeys, which are not used extensively in teratological testing. The data showing that the *R*-enantiomer has more of a sedative effect and the *S*-enantiomer more of an immunomodulating and teratogenic effect may give some hope for an understanding of mechanisms of action, but it does not help much to use pure enantiomers since racemisation occurs in the blood.

The come-back of thalidomide in leprosy has resulted in at least 1000 new cases of malformed babies, in particular in Latin America. The teratogenicity of thalidomide presently poses a similar problem for the regulatory agencies in Europe.

Nonsteroidal anti-inflammatory drugs (NSAIDs)

It has long been recommended that NSAIDs should not be used late in pregnancy owing to known effects during parturition, including bleeding, premature closure of the ductus arteriosus, disturbed kidney function leading to oligohydramnios, and inhibition of uterine contractions. More recent studies have shown that NSAIDs, including COX2 inhibitors, may disturb ovulation, implantation of the blastocyst and decidualisation, most likely because prostaglandins are involved in several of these inflammatory-like processes. It is also known that mice carrying deletions of the gene encoding COX2 have decreased ovulation and fertilisation rates, failure of implantation and incomplete decidualisation. More recent epidemiological reports indicate a nearly twofold increase in minor heart malformations upon exposure to NSAIDs in early pregnancy.

Recreational drugs

Children of alcoholic mothers may be born with morphological as well as functional disturbances. There are three components of a fully developed *fetal alcohol syndrome.*

*   Prenatal and/or postnatal growth retardation (weight and/or length)
*   Aberrations in the central nervous system, such as microcephaly, psychomotor retardation and neurological, learning and behavioural disturbances
*   A characteristic facial expression, including short eye fissure, broad, flat nasal bridge, underdeveloped midface, long upper lip and diffuse philtrum

Cocaine may cause central nervous system irritation, cardiac anomalies, apnoea and feeding difficulties in the offspring. Decreased birth weight, body length and head circumference revealed growth retardation in cocaine-exposed infants during the last months of gestation. Other observations include abnormal auditory brainstem responses among infants with prenatal cocaine exposure, and deficits in language functioning.

It is well established that *tobacco smoking* can cause lower birth weight, increased perinatal death and an increased incidence of deficiencies in attention, motor control and perception (DAMP).

Conclusion

It is clear that exposure to medicinal drugs as well as recreational drugs and other chemicals during development may be detrimental to the offspring. It is important to test properly for such effects preclinically, since they are irreversible and it takes a long time after introduction of a drug to the market before there are enough cases of women exposed during pregnancy to exclude or verify an effect on the offspring. This situation also calls for better understanding of mechanisms of action of teratogens, as well as methods for studying such effects. Today there is hope that embryonic stem cells may contribute to such a methodological improvement, but there is still much research to be done to show that such cells would mirror embryonic developmental processes. Another approach that seems promising is to focus on more physiological parameters such as the haemodynamics of the offspring during specific periods of development. It is becoming increasingly clear that the pharmacological mechanisms of a drug apply in the embryo as well, although downstream effects may differ from those in the adult, and may be detrimental to the embryo.

Further reading

Carney EW, Scialli AR, Watsom RE, DeSotto JM (2004). Mechanisms regulating toxicant disposition to the embryo during early pregnancy: an interspecies comparison. *Birth Defects Res C Embryo Today* 72: 345–360.

Danielsson BR, Skold AC, Azarbayjani F (2001). Class III antiarrhythmics and phenytoin: teratogenicity due to embryonic cardiac dysrhythmia and re-oxygenation damage. *Curr Pharm Des* 7: 787–802.

Danielsson BR, Johansson A, Danielsson C, Azarbayjani F, Blomgren B, Skold AC (2005). Phenytoin teratogenicity: hypoxia marker and effects on embryonic heart rhythm suggest an hERG-related mechanism. *Birth Defects Res A Clin Mol Teratol* 73: 146–153.

Kallen BA. (2005). Methodological issues in the epidemiological study of the teratogenicity of drugs. *Congenit Anom (Kyoto)* 45: 44–51.

Klaasen CD, ed. (2001). *Casarett & Doull's Toxicology. The Basic Science of Poisons*, 6th edn. New York: McGraw-Hill.

Mark M, Ghyselinck NB, Chambon P (2004). Retinoic acid signalling in the development of branchial arches. *Curr Opin Genet Develop* 14: 591–598.

Webster WS, Freeman JA (2001). Is this drug safe in pregnancy? *Reprod Toxicol* 15(6): 619–629.

5

Genotoxicity

Björn Hellman

Genotoxic agents are chemicals, physical factors and biological agents that can induce different types of alterations of the genetic material, both permanent changes (i.e. different types of mutations) and modifications that can be repaired (premutagenic lesions, primary DNA damage, inactivated DNA). Focusing on chemically induced genotoxicity, this chapter will discuss some fundamental principles of genetic toxicology including different 'end-points' that are screened for when evaluating the potential genotoxicity of drugs, the mechanisms behind different types of genetic alterations, and some of the test systems used when screening for genotoxicity. Since a reader of this book should be familiar with some of the difficulties that may arise when managing a situation in which patients are exposed to genotoxic drugs, the chapter also includes a small dose of safety evaluation.

A permanent change in the genetic material alone is not an adverse effect unless it leads to a biologically significant change in the phenotype that might be quite deleterious, both for the exposed individuals and for their offspring. If mutations occur in somatic cells of an exposed individual, this may increase the risk of tumour development, especially if the mutations lead to activation of proto-oncogenes or inactivation of tumour suppressor genes. Proto-oncogenes are involved in normal cellular growth and differentiation, and they are usually inactivated in mature somatic cells. A critical mutation in such a gene may activate the proto-oncogene so that it becomes an oncogene, which in turn leads to overexpression of the growth-stimulatory activity in the cell. Tumour suppressor genes are active in mature somatic cells, contributing to different checkpoint control functions that ensure that the DNA and the chromosomes are intact, and that critical stages of the cell cycle are completed before the following stage is initiated. If critical mutations occur in fetal somatic cells, this may lead to an increased risk of congenital abnormalities. Mutations in the germ cells of future parents may increase the risk of different types of genetic diseases in their offspring and later generations. Moreover, owing to their mechanism of action, it is often assumed that genotoxic agents may act without a threshold dose.

Genotoxicants, mutagens, clastogens and aneugens

A genotoxic agent was defined above as a chemical, physical or biological factor that could induce different types of reversible and irreversible changes in the genetic material. This is a broad definition, and some authors define a genotoxic chemical as a DNA-reactive compound that induces DNA damage by interfering directly with the DNA molecule. A chemical mutagen is then referred to as an agent that induces mutations at the level of the DNA molecule. Agents that provoke different types of stable or unstable morphological and structural chromosomal aberrations are usually called clastogens, and compounds that give rise to an abnormal number of chromosomes are often referred to as aneugens. Since primary DNA damage may lead to both gene mutations and chromosomal aberrations, it is not surprising that DNA-reactive compounds are usually found to be DNA-damaging, mutagenic and clastogenic. Since an abnormal number of chromosomes usually follows from the failure of homologous chromosomes or sister chromatids to separate during mitosis or meiosis without any direct damage to the genetic material, DNA reactivity is not necessarily a prerequisite for aneugenicity.

DNA-reactive compounds and chemical mutagens, clastogens and aneugens can act either by themselves (as parent compounds) or after biotransformation ('bioactivation'). Consequently, as for many other 'toxicological end-points', there are several important general aspects that should be addressed when evaluating the potential genotoxicity of chemicals. Among these are the exposure conditions, the bioavailability and toxicokinetics of the parent compound, as well as the balance between metabolic bioactivation and detoxification (most DNA-reactive compounds need bioactivation before they can interact with the DNA). However, there are additional aspects that are of particular importance when it comes to the genotoxicity of chemicals. Among those are the nature of the DNA interaction for DNA-reactive compounds, the phase of the cell cycle and DNA replication at which the insult occurs, and the fidelity of the DNA repair (probably one of the most important factors in genetic toxicology).

End-points for evaluating potential genotoxicity

As shown in Table 5.1, genotoxicity includes many different types of 'genetic end-points', including both premutagenic lesions that can be repaired and different types of permanent changes of the genetic information that are inherited from one cell generation to the next (mutations). DNA adducts are chemical modifications of DNA resulting from the covalent binding of a reactive electrophilic chemical species with nucleophilic sites in DNA (Figure 5.1). If the DNA adducts and the primary DNA lesions are not correctly repaired before DNA replication, some of these adducts will lead to base-pair substitutions, deletions, insertions, recombinations, and other replication errors. These errors will become permanent genetic alterations once the cell undergoes mitosis (somatic cells) or meiosis (germ cells).

Table 5.1 A representative sample of genetic end-points (and techniques) that can be evaluated when screening for the potential genotoxicity of chemicals

Direct interaction with DNA (as shown in, e.g., binding studies *in vivo* or *in vitro* using radiolabelled compounds or the ^{32}P-postlabelling assay)	• Alkylated DNA • Bulky DNA adducts
Primary DNA-damage (as shown, directly or indirectly in, e.g., *in vivo* or *in vitro* versions of the HPC/DNA-repair test; the comet assay or alkaline elution assays)	• AP-sites (apyrimidininic or apurinic sites) • DNA–DNA cross-links (interstrand or intrastrand cross-links) • DNA–protein cross-links • Pyrimidine dimers • Intercalations • DNA strand breaks (single-strand or double-strand breaks)
Gene mutations (as shown in, e.g., bacterial reverse mutation assays, mammalian *in vitro* forward mutation assays, or *in vivo* assays for gene mutations in somatic cells and germ cells)	• Base substitutions (transitions or transversions) • Frameshifts (i.e. deletions or additions of individual base pairs in a coding region)
Structural chromosomal aberrations (as shown *in vivo* or *in vitro* in mammalian cytogenetic assays, e.g., in the micronucleus test or by conventional metaphase analysis)	• Chromatid and chromosome breaks • Translocations

HPC, hepatocyte primary culture.

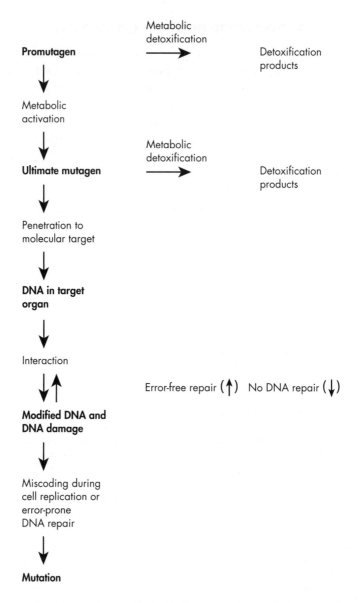

Figure 5.1 A common picture that holds for many chemical mutagens, but not all.

The central dogma of molecular biology and genetic toxicology is that information flows from DNA to RNA (transcription) and from RNA to proteins (translation), and that the genetic information flows between different generations of cells via mitosis and meiosis. This means that permanent changes in DNA can be studied either directly using, for example, Southern blotting or the polymerase chain reaction (PCR) technique, or, as in most short-term tests for genotoxicity, indirectly by monitoring altered phenotypic expression.

Whereas DNA adducts and different types of primary DNA damage are increasingly used as biological markers of exposure in epidemiological studies of environmental exposures to genotoxic agents, the main focus when testing for the potential genotoxicity of drugs and other chemicals is still on different types of mutations. By tradition, mutations are usually divided into the following three major groups: (i) gene mutations; (ii) larger-scale structural chromosomal mutations; and (iii) numerical chromosomal changes. Among these, gene mutations and chromosomal aberrations seem to be of primary concern when testing for the genotoxicity of drugs.

Aneuploidy (gain or loss of one or more chromosomes) is no doubt an important cause of genetic disease, but this genetic end-point is often more or less neglected in conventional genotoxicity testing. One reason for this may be the absence of validated test procedures; another that the mechanisms behind an erroneous chromosome number most likely exhibit a threshold dose. Nevertheless, it should be emphasised that information on aneuploidy may be derived from some of the tests used when screening for clastogenic effects. It should also be mentioned that other types of genetic events that are not easily categorised as being typical gene mutations or structural chromosomal mutations can be monitored when evaluating the potential genotoxicity of drugs. These include different types of recombinations (e.g. gene conversion, reciprocal exchanges between homologous chromosomes and sister chromatid exchanges), gene amplifications and insertion mutations.

DNA adducts and primary DNA damage

DNA adducts are increasingly used as 'molecular biomarkers' of exposure in epidemiological studies, but occasionally they have also been used to assess the genotoxic potential of chemicals. The biological significance of the DNA adducts must be assessed on the basis of both adduct heterogeneity and of the cell and tissue specificity for adduct formation, persistence and repair. When monitoring DNA adducts, it is important that the duration and timing of the exposure is known for a proper evaluation of the biological significance of a given adduct concentration. Some adducts result in mutations, others do not. Some DNA adducts are repaired quickly, others hardly at all. Sensitive techniques based on physicochemical or immunological methods have been developed for the detection of various types of DNA adducts. The most frequently used are the ^{32}P-postlabelling method, various immunoassays, GC-MS techniques (gas chromatography coupled with mass spectrometry), and synchronous fluorescence spectroscopy. Each of these techniques has its advantages and disadvantages.

DNA adducts, bulky or not, can be looked upon as structural modifications of the DNA molecule and therefore also as potential premutagenic lesions. There are also other types of changes in and/or interactions with DNA that may increase the risk of gene mutations and/or structural chromosomal mutations if they are not repaired correctly before replication and cell division: dimers, cross-linking (DNA–DNA and DNA–protein cross-links), intercalation, apurinic and apyrimidinic sites (AP-sites), different types of oxidative DNA damage, and DNA strand breaks (single-strand and double-strand breaks). It is beyond the scope of the present chapter to discuss the different types of primary DNA damage in detail, but since some of these premutagenic lesions also occur naturally in cells, DNA strand breaks and oxidative DNA damage will be briefly discussed.

DNA single-strand breaks, and to some extent DNA double-strand breaks, are typical representatives of DNA damage that are both naturally occurring and induced by genotoxic carcinogens. Several techniques have been used to monitor DNA strand breaks both *in vitro* and *in vivo* (e.g. the alkaline elution technique, the nick translation assay, and neutral or alkaline single-cell gel electrophoresis). Being a relatively sensitive, simple and rapid technique, the alkaline version of the single-cell gel electrophoresis assay (also known as the 'comet assay') is increasingly used for the detection of DNA single-strand breaks. Because of its unique design, allowing the measurement of damage in individual cells, it is possible to determine whether all cells show the same degree of damage, or whether there is a heterogeneous response to the genotoxic insult of interest.

Although still not regularly monitored for when screening for the potential genotoxicity of a candidate substance in the early phase of drug development, oxidative DNA damage (which can be induced by both endogenous and environmentally induced reactive oxygen species) has been the focus of interest during recent years, both as an exposure marker and as an effect marker in experimental and epidemiological studies of various diseases (including cancer). One of the most important molecular biomarkers for oxidative DNA damage is 8-oxy-deoxyguanosine, which can be measured by HPLC (high-performance liquid chromatography) with electrochemical detection and by other techniques (including modified versions of the comet assay).

Gene mutations

Current gene mutation assays usually select for a change or a loss of a normal protein produced by specific genes. A permanent change of the genetic code at the level of the DNA molecule is usually referred to as a gene mutation, but this type of mutation can also be referred to as a point mutation (the DNA sequence change affects only one or two base pairs) or even a block mutation (the change of the DNA sequence affects a relatively large number of base pairs, possibly involving several different genes). Nevertheless, by definition, all permanent changes in a DNA sequence will lead to an altered genotype, but only some of these changes will also lead to biologically significant phenotypic alterations that can be selected for and monitored.

It is generally stated that permanent changes of the genetic code at the level of the DNA molecule are due to (i) base-pair substitutions or (ii) frameshifts (additions or deletions of a limited number – not in multiples of three – of individual base pairs in the DNA sequence). The base-pair substitutions are either transitions or transversions. A transition occurs when a pyrimidine is replaced by another pyrimidine or when a purine is replaced by another purine. In a transversion a pyrimidine is replaced by a purine or vice versa. An alternative way of classifying different types of gene mutations is based on whether they affect the translation of information from DNA to RNA. A given base-pair substitution may not affect the transcription at all, or it may lead to erroneous transcription (which may or may not be deleterious for the cell). A gene mutation may also lead to a complete stop of the transcription (see Table 5.2).

Gene mutations can be classified in still other ways. It is well known, for example, that gene mutations are usually either recessive (not expressed in a heterozygous state) or dominant (expressed even in a heterozygous state). Moreover, whereas some gene mutations, in tumour suppressor genes, for example, might be rather critical in the multistage process of tumour development, other gene mutations can be completely silent during the entire lifespan of the cell carrying the mutation. However, a silent mutation located in a region of the DNA that is not expressed might become critical if it is relocated to another region (e.g. by a chromosomal translocation). The occurrence of a lethal mutation leading to the death of a single cell will in most cases not lead

Table 5.2 Some point mutations will affect the flow of genetic information, others will not

DNA		RNA		Amino acid
AAA No mutation (Wild type)	→	UUU	→	Phenylalanine
AAG A point mutation (Transition)	→	UUC	→	Phenylalanine No change in phenotype (Neutral mutation)
AGA A point mutation (Transition)	→	UCU	→	Serine Wrong amino acid is formed (Missense mutation)
ATT A point mutation (Frameshift)	→	UAA	→	Stop codon No amino acid is formed (Nonsense mutation)

A, adenine; C, cytosine; G, guanine; T, thymine; U, uracil.

to any serious consequences (at least not in fully differentiated somatic cells).

Some of the assays used when screening for gene mutations (especially those based on prokaryotic organisms) are so-called reverse mutation assays. In these assays (e.g. the Ames test, see below), the marker gene carries a mutation that can be reverted to wild type by a new mutation, a so-called back mutation. However, most assays for gene mutations (e.g. the mouse lymphoma L5178Y TK-locus assay, see below) are forward mutation assays. These tests detect mutations from the parental type to the mutant form which give rise to a change in an enzymatic or functional protein.

Most short-term tests used when screening for chemically induced gene mutations monitor the mutation frequency in a specific marker (indicator) gene, but another, more recent approach has been to monitor also the mutational spectra in these genes (e.g. in the X-linked HPRT (hypoxanthine-guanine phosphoribosyl-transferase) locus). The location and type of point mutations in a specific sequence of nucleotides define a mutational spectrum, and promising attempts have been made to associate specific chemical exposures with specific mutational spectra. An alternative approach for the measurement of chemically induced gene mutations that does not require a prior selection of mutant cell populations is based on the use of the restriction site mutation technique. This technique, which is commonly used in molecular biology, is based on the detection of DNA sequences that are resistant to the cutting action of specific restriction enzymes. When the resistant sequences have been found, they can be amplified using the PCR technique and then sequenced so that any potential mutation can be detected.

The explosive expansion of molecular biology has, without doubt, expanded the field of toxicogenomics and proteomics enormously during the last few years, but many of the techniques that are used to study the expression of thousands of genes at one time (by using microarrays) are still too expensive to be used for the routine testing of the potential genotoxicity of chemicals. Another problem is that many of the new techniques and protocols used have not been sufficiently validated, at least not when it comes to their usefulness for risk assessment/safety evaluation of genotoxic agents.

Structural and numerical chromosomal aberrations

Structural and numerical chromosomal aberrations represent two different types of chromosomal mutations. Whereas numerical chromosomal mutations follow from the failure of homologous chromosomes or sister chromatids to separate during meiosis or mitosis (this phenomenon is called 'non-disjunction' and does not involve any direct damage to the genetic material), structural chromosomal aberrations probably arise as a result of erroneous repair of DNA damage in the G_0 phase of the cell cycle.

Typical structural chromosome aberrations (including, for example, acentric fragments, dicentric chromosomes, ring chromosomes, chromatid breaks, internal rearrangements within a chromosome and exchanges of chromosomal material btween chromosomes) are monitored using either classical cytogenetic methods or the micronucleus test (see below), but balanced translocations and inversions can be difficult to monitor without a so-called banding analysis. The classical cytogenetic analysis of chromosome structures in metaphase cells using conventional staining techniques and light microscopy is rather laborious and requires a highly trained eye to evaluate and classify the different types of changes that can be observed. A modern technique, fluorescence in-situ hybridisation (FISH), is a molecular cytogenetic technique in which fluorescently labelled small DNA or RNA probes are hybridised to chromosomes on interphase or metaphase spreads on slides. The FISH technique makes it possible to visualise specific genes and thereby also to define their number and chromosomal localisation.

Depending on their ability to persist in dividing cell populations, the structural chromosomal aberrations are either balanced (stable) or unbalanced. Most unstable aberrations (acentric fragments, ring chromosomes and other asymmetrical rearrangements) will lead to cell death and therefore cannot be classified as mutations. In contrast, many stable chromosomal aberrations (e.g. balanced translocations and other symmetrical rearrangements) can be transmitted to the next cell generation, and these can therefore be looked upon as chromosomal mutations that may have great biological significance in, for example, the multistage process of tumour development.

Short-term tests for genotoxicity

The potential genotoxicity of drugs is usually tested in a battery of short-term tests including different types of genetic end-points (typically gene mutations and structural chromosomal aberrations, and sometimes also primary DNA damage). Most of the short-term tests are relatively easy to perform, are not very expensive, and usually follow standardised and validated test protocols such as those given in the OECD (Organisation for Economic Co-operation and Development) Guidelines for the Testing of Chemicals. It must be emphasised that the results obtained from a screening of potential genotoxicity, even in a battery of well-established short-term tests, permit only *qualitative* judgements about the potential human hazard. It is consequently not possible to make any quantitative judgements about, say, a potential cancer risk that might follow from an exposure to an agent that was found to be genotoxic in a short-term test.

Since most genotoxic agents act through one or several reactive intermediates, most *in vitro* assays for genotoxicity would be useless if they did not employ a metabolic activation system. What is known as an S9 fraction is therefore used in most of these tests, especially in those based on prokaryotic organisms but also in many assays based on mammalian cells. The S9 (9000g fraction) is usually prepared from a liver homogenate from adult male rats, pretreated with an inducer of cytochrome P450. The most commonly used inducer is a mixture of polychlorinated biphenyls (Aroclor 1254). The (unavoidable) use of S9 in many *in vitro* tests makes them more sensitive towards DNA-reactive metabolites, but the high enzyme activity and the balance between phase 1 and phase 2 enzymes in these assays does not (at all) reflect the normal situation in human cells. Moreover, bioactivation by conjugation is not detected in the S9 because the necessary co-factors are not present. There is also some evidence suggesting that the S9 may itself be slightly toxic to the indicator cells.

A typical short-term test for genotoxicity should employ at least one negative control (usually the vehicle), and one or two positive controls (compounds that are known to be genotoxic). Typical positive controls include the direct-acting mutagen ethyl methanesulfonate when testing without a metabolic activation system, and 3-methylcholanthrene when an exogenous metabolic activation system is used in the assay.

In order to be able to see any dose–response relationships (an important aspect in genetic toxicology also), a short-term test for genotoxicity should employ multiple concentrations (doses) of the test compound (at least three different concentrations). These should be based on a preliminary dose selection test, and the highest dose/concentration should then (if possible) be toxic for the indicator cells (often manifested as reduced cell viability). It goes without saying that compounds that induce DNA damage, chromosomal aberrations and/or an increased mutation frequency at low, non-toxic, concentrations are more problematic than those that are genotoxic only at high and clearly cytotoxic concentrations. A common feature when testing the potential genotoxicity of chemicals is also that the results from each individual assay should be confirmed in at least one independent experiment (often two).

In 1997, the European Agency for the Evaluation of Medicinal Products (EMEA) recommended the following standard test battery for genotoxicity testing of pharmaceutical drugs: (i) a test for gene mutations in bacteria; (ii) either an *in vitro* test with cytogenetic evaluation of chromosomal damage using mammalian cells, or an *in vitro* mouse lymphoma TK-locus assay; and (iii) an *in vivo* test for chromosomal damage using rodent haematopoietic cells. For compounds giving negative results (no gene mutations or chromosomal aberrations), the completion of the suggested standard test battery (performed as indicated in various guidelines) was considered to provide a sufficient level of safety to demonstrate the absence of genotoxicity. For compounds giving positive results it could, depending on their therapeutic use, be necessary to perform more extensive testing (including other genetic end-points such as DNA strand breaks or increased DNA repair activity). There follows below a brief description of four different tests that are used worldwide when screening for the potential genotoxicity of pharmaceutical drugs and other chemicals.

The HPC/DNA-repair test (unscheduled DNA synthesis in mammalian cells)

Unscheduled DNA synthesis (UDS) is an indication of DNA repair, and increased DNA repair typically follows from different types of DNA damage. Increased UDS is monitored as an increased incorporation of tritium-labelled thymidine ([³H]thymidine, a selective DNA precursor) into the DNA. The cells must not be in the S-phase of the cell cycle because, during this phase, the overwhelming part of the thymidine incorporation would be due to DNA replication. The incorporation of [³H]thymidine can be monitored by liquid scintillation counting or microautoradiography, and this can be done in many different types of mammalian cells.

The use of primary cultures of hepatocytes (typically from rats) does not require the use of S9, which probably explains why the HPC (hepatocyte primary culture) DNA-repair test is preferred by many laboratories when monitoring UDS.

The *Salmonella* mammalian-microsome mutation assay (Ames test)

The Ames test is an *in vitro* assay based on previously mutated micro-organisms and it measures his⁻ to his⁺ reversions (back mutations). The *Salmonella typhimurium* bacteria are exposed to different concentrations of the test compound, both with and without S9, and after a suitable period of incubation in minimal medium, revertant colonies are counted and compared with the number of spontaneous revertant colonies in the negative control. Several different strains of *S. typhimurium* (e.g. TA98 and TA100) with different susceptibilities towards different types of mutagens are used when assessing the potential mutagenicity of a compound in the Ames test. A common feature of the different strains is that they are mutated not only in the histidine gene (i.e. the indicator gene), but also in other genes. Their sensitivity towards mutagenic agents has been increased by a mutation increasing the permeability of their cell walls (deficient in lipopolysaccharide) and by another mutation that reduces their capacity for DNA excision repair (a very important defence mechanism in our cells).

The mouse lymphoma L5178Y/TK-locus assay

The major objective of the mouse lymphoma assay is to monitor forward mutations in the thymidine kinase (TK) locus, but mutant colonies can also be the result of both structural chromosomal changes (leading to a loss of the TK gene) or numerical chromosomal changes (leading to a loss of the chromosome carrying the TK gene). The assay is based on mouse lymphoma L5178Y cells, which are heterozygous at the TK-locus (TK$^{+/-}$). A gene mutation in the TK-locus can change the genotype so that it becomes homozygous. Both TK$^{+/-}$ and TK$^{-/-}$ cells can grow in normal medium, but since TK$^{-/-}$ cells lack the important salvage enzyme thymidine kinase, only TK$^{-/-}$ cells can grow in a medium containing the cytotoxic pyrimidine analogue 5-bromo-2′-deoxyuridine (BrdU). Owing to the lack of the salvage enzyme, the TK$^{-/-}$ cells will not incorporate the antimetabolite BrdU, and all nucleotides needed for cellular metabolism in TK$^{-/-}$ cells have to be produced by *de novo* synthesis (which is independent of thymidine kinase). Consequently, the genotoxicity of a test compound in the mouse lymphoma TK-locus test is determined by comparing the growth of control cells and exposed cells, with and without BrdU in the medium, both in the presence and in the absence of S9.

The micronucleus test *in vivo*

The micronucleus test is a mammalian *in vivo* test that can detect both structural chromosomal damage (i.e. clastogenic effects) and damage to the mitotic apparatus (i.e. numerical chromosomal aberrations). The micronucleus assay is attractive because it is much easier to perform than the classical cytogenetic tests where the chromosome aberrations are typically classified and scored in stained cells that have been arrested in metaphase. In a typical micronucleus test, mice that have been exposed to different doses of the test compound (usually two times within 24 hours) are killed 6 hours after the last treatment. Bone marrow cells are collected from the femurs; smears are prepared and stained, and then analysed by light microscopy (scoring polychromatic erythrocytes for micronuclei).

Polychromatic erythrocytes (PCEs) are used because they do not carry a nucleus (it is extruded when an erythroblast develops into an erythrocyte). This makes it easier to find any micronucleus remaining in the cytoplasm of the PCE. Micronuclei are small particles constituting either acentric fragments of chromosomes or entire chromosomes left behind during the anaphase stage of cell division. At least 1000 PCEs per animal should be analysed for the incidence of micronuclei, both in controls and treated animals. The ratio of PCEs to normochromatic erythrocytes (i.e. fully mature erythrocytes; NCEs) should also be established by counting at least 1000 NCEs per animal (usually 5 animals/treatment). The latter is done in order to capture any potential cytotoxicity of the test compound.

An important issue in the safety evaluation of drugs: threshold doses or not?

A direct interaction between a DNA-reactive agent and DNA is only one of several pathways that may lead to a permanent change of the genetic information. A mutation may also follow from other, more indirect events (e.g. erroneous DNA repair or an unbalanced precursor pool). An important question is whether it is reasonable to assume, by default, that chemically induced mutations occur at random (i.e. are the result of a stochastic event without a threshold dose). Clearly, DNA adducts and chemically induced DNA damage may lead to both gene mutations and chromosomal aberrations. As a worst-case assumption, it is then often assumed that this is a non-threshold event. However, thresholds may actually be involved in such a case, especially for compounds that must be bioactivated to form DNA-reactive intermediates, and where low concentrations of these reactive intermediates are taken care of by various cellular defence systems such as glutathione (see below).

Errors during the DNA replication and erroneous DNA repair may also lead to both gene mutations and chromosomal aberrations; although in most cases it seems fair to assume that a threshold dose exist for these events, these errors may possibly also occur at random (the DNA polymerases involved both in the replication and in many repair processes are not absolutely faithful in their replication of the template DNA strand). Gene mutations and structural chromosomal aberrations may also be due to an unbalanced precursor pool, but in this case it is hard to believe that a non-threshold dose event is possible. Another threshold dose event leading to primary DNA damage (and therefore possibly also to gene mutations and chromosomal aberrations) is cytotoxicity. Non-disjunction is also a clear threshold dose phenomenon (possibly leading to a numerical chromosomal mutation).

The issue of the existence of threshold doses can be illustrated using paracetamol (acetaminophen) as an example. It has been shown that high concentrations of this compound cause chromosomal damage, both *in vitro* and *in vivo*. It has also been shown that NAPQI (*N*-acetyl-*p*-quinone imine; a reactive metabolite of paracetamol) can bind covalently to DNA and induce primary DNA damage (but, strangely enough, apparently not gene mutations). If paracetamol were an environmental contaminant like benzene (which actually has a rather similar genotoxicity profile to that of paracetamol), a regulatory agency would probably follow the so-called precautionary principle and suggest that this agent might be genotoxic without a threshold dose.

In the case of paracetamol, experts found that all available data (including additional information about the toxicological profile of paracetamol) pointed to three possible mechanisms of paracetamol-induced genotoxicity: (i) inhibition of ribonucleotide reductase; (ii) increase in cytosolic and intranuclear Ca^{2+} levels, and/or (iii) DNA damage (see *Mutat Res.* (1996) 349: 263–288). All these effects were considered to follow a threshold dose model. The formation of NAPQI is mediated by cytochrome P450, and since the covalent binding of NAPQI to DNA and the subsequent DNA damage occur only after depletion of cellular glutathione, it seems logical to assume a threshold dose for this event. The paracetamol-induced inactivation of ribonucleotide reductase (probably via a free-radical phenoxy species that is formed independently of cytochrome P450) leads to an unbalanced precursor pool (affecting both the replication and DNA repair); a typical threshold dose phenomenon. High doses of paracetamol are cytotoxic and are associated with increased intranuclear levels of Ca^{2+}. The increase in calcium concentration leads to an activation of endonucleases and thereby also to DNA fragmentation.

High-fidelity DNA repair protection

The DNA repair capacity represents an important parameter when it comes to an individual's susceptibility towards genotoxic agents. Our cells are normally rather well protected against mutations resulting from different types of DNA modifications, and mammalian cells have evolved several different DNA-repair pathways. The most important ones are obviously those that can be classified as being error-free. The three major groups of error-free repair pathways in mammalian cells are: (i) direct repair by reversal (e.g. by dealkylation); (ii) mismatch repair; and (iii) two different types of excision repair (base excision repair and nucleotide excision repair). It is beyond the scope of the present chapter to describe the different pathways of DNA repair in detail, but the basic principles underlying most (but not all) repair processes are

(i) recognition of damage; (ii) removal of damage; (iii) DNA repair synthesis; and (iv) ligation of DNA. In the error-free repair pathways all these events should normally occur before replication and cell division. In our cells, we also have a 'last chance' repair (DNA double-strand break repair by, e.g. postreplicative recombinations), but this type of repair is unfortunately of low fidelity and therefore error-prone (the integrity of the DNA is improved but still altered).

Genotoxic agents and genetic disease

It is well known that many human diseases are associated with different types of mutations; it is also well known that a number of chemicals can damage the genetic material in the germ cells of experimental animals and, by doing so, also affect the offspring of these animals. It therefore seems logical to assume that genotoxic chemicals can also increase the risk for genetic disease by germ cell mutations also in humans. An intellectual dilemma is that genetic diseases or abnormalities resulting from chemically induced mutations in human germ cells have not yet been detected. There can be several different reasons for this and, although not discussed here, it should be emphasised that epidemiological studies of heritable risks following from exposures to genotoxic agents are fraught with various methodological problems. When it comes to drug-induced malformations in humans, most seem to be the result of the pharmacological action of the drugs and not of mutations (see Chapter 4).

Several assays are available for monitoring chemically induced mutations in germ cells of experimental animals. Among these are the specific-locus test (which measures recessive gene mutations); the translocation test (which measures reciprocal mutations), and the dominant-lethal test (which measures structural and numerical chromosomal mutations). In general, these tests are quite expensive and time-consuming and require many animals. This is a problem, considering that a complete assessment of the mutagenic activity of a chemical in germ cells of mammals should ideally include all germ cell stages, both in males and females.

Conclusion

Given that mutagenic agents may increase the risk for both malignant diseases and heritable defects at all doses above zero (at least theoretically), it is not surprising that screening tests for genotoxicity have become one of the primary means for identifying potentially hazardous chemicals. This seems to be particularly true when it comes to the lead candidate selection stage of drug development in pharmaceutical companies, where the outcome in the short-term tests for genotoxicity requiring relatively small amounts of the candidate substance (an important aspect in the early phase of drug development) provides valuable information about the potential toxicological profile of the candidate drug before it enters the more extensive, extremely costly and time-consuming regulatory toxicity testing required for registration.

Further reading

Allen JW, Ehling UH, Moore MM, Lewis SE (1995). Germ line specific factors in chemical mutagenesis. *Mutat Res.* 330: 219–231.

Li AP, Heflich RH, eds (1991). *Genetic Toxicology: A Treatise.* Boca Raton: CRC Press.

McGregor DB, Price JM, Venitt, eds (1999). *The Use of Short-term and Medium-term Tests for Carcinogens and Data on Genetic Effects in Carcinogenic Hazard Evaluation,* IARC Sci. Publ. No. 146. Lyon: IARC.

6

Carcinogenicity of drugs

Gerard J Mulder

Cancer is a major cause of death in the Western world: approximately 25% of the population will develop cancer of some type during their lifetime, and approximately 20% of the population die as a consequence of cancer. The incidence of cancer at different sites varies greatly between males and females as well as between countries: while breast cancer is prominent in women (15–20% of cancer mortality), the prostate is a major cancer site in men (10–20%). The WHO (World Health Organization) mortality database (see the IARC 'Cancer *Mondial*' database at www-dep.iarc.fr) gives relevant data for countries worldwide. Genetic, economic and lifestyle differences play a major role in the incidence of cancer at various sites. Some causes of human cancer are well known, such as smoking or asbestos exposure; also some drugs may cause cancer (Table 6.1). However, in most cases the cause of cancer in an individual is not known. Increasingly, evidence for genetic predisposition to develop cancer is found, for instance for breast cancer.

In this chapter the terms 'cancer', 'carcinogenic' and 'tumour' will be used to indicate processes in which neoplastic growth is involved, both malignant and benign. The famous cancer researcher Henry Pitot defined a neoplasm as follows: 'A neoplasm is a relatively autonomous growth of tissue.' It is 'relatively autonomous' because neoplastic growth is not regulated by the normally operating processes that tend to limit growth at a certain rate or stage. However, at the same time the neoplasm is not totally autonomous: it still depends on outside factors such as nutrients and growth factors. The component 'growth' indicates that the neoplasm grows and shows cell division, but this is not necessarily a rapid growth. Growth may be slow at one time for a particular tumour while for another tumour it may be extremely rapid. Thus, certain prostate tumours may be dormant for the whole lifespan of an individual, because they do not grow sufficiently rapidly to cause problems.

A neoplasm arises in a given tissue and originates from a single

Table 6.1 Classification of some drugs in groups 1, 2A and 2B by IARC

Group 1: Proven carcinogenic to humans
Azathioprine
Chlorambucil
Ciclosporin
Diethylstilbestrol
Etoposide
Herbal remedies containing plant species of the genus *Aristolochia*
Melphalan
8-Methoxypsoralen
MOPP and other combined chemotherapy including alkylating agents
Oestrogen therapy, postmenopausal
Oral contraceptives, sequential
Tamoxifen
Thiotepa
Analgesic mixtures containing phenacetin

Group 2A: Probably carcinogenic to humans
Anabolic steroids
Chloramphenicol
Cisplatin
Doxorubicin (adriamycin)
Etoposide
Phenacetin

Group 2B: Possibly carcinogenic to humans
Bleomycins
Griseofulvin
Mitomycin
Oxazepam
Phenobarbital
Phenolphthalein
Phenytoin
Zidovudine

MOPP, chlormethine (mustine), vincristine (Oncovin), procarbaxine, prednisolone.

cell. It therefore often retains many (but not all) characteristics of that original cell type, albeit that during its outgrowth the tumour is usually subject to major changes as a consequence of mutations within the cells of the tumour. Malignant neoplasms are usually genetically very labile, so that further mutated cells can continuously arise within a single tumour. This lies behind the drug resistance often observed after the initial therapeutic response of a tumour to a cytostatic drug: the mutant cells that are resistant are being selected because of their higher survival potential.

Benign and malignant tumours and their classification

A major distinction is made between benign and malignant tumours. Benign tumours are those that are local (usually surrounded by connective tissue), do not show invasive behaviour and do not metastasise (Figure 6.1). In contrast, malignant tumours do metastasise from their original site to one or more other organs. Pathologists can often assign a tumour to one of these two classes (e.g. a benign 'hepatocellular adenoma' or a malignant 'hepatocellular carcinoma'), but frequently there are tumour types in certain tissues that are very hard to classify, leading to mistaken classification. Because there are differences between human and animal tumours, their classification on the basis of microscopic morphology requires much experience in specialised human or animal pathologists. Classification can be improved by immunostaining of proteins characteristic of a certain tumour or tissue type.

To assess the carcinogenic effects of a chemical in animal experiments, the pathologist has to evaluate (tissue slices of) a full autopsy of all the treated and control animals in order to detect, classify and quantitate all chemically induced tumours in that species. Of particular significance is their classification as malignant or benign. Obviously very small tumours (i.e. early stages) can easily be missed. It should be mentioned that this is done after a period of up to two years of exposure (see below). For drugs, these studies are combined with chronic, general toxicity studies.

The International Agency for Research on Cancer (IARC) in Lyon (France) has defined a carcinogen as an agent that is capable of increasing the incidence of *malignant* tumours. By this definition, benign tumours do not count. Benign tumours are always quantified and reported, but they will only be added to malignant tumours if they appear to represent a stage in the progression of a tumour to malignancy. For instance, if the liver contains both hepatocellular carcinomas (malignant) and adenomas (benign), these might be added up for a dose–response relationship. A chemical that induces only benign tumours in animal experiments is not considered a carcinogen according to the IARC definition. Of course, a benign tumour at a site where its surgical removal is impossible may still be lethal for the individual. For evaluation of the carcinogenic potential of drugs, the benign or malignant character is taken into consideration. Drugs for which evidence is available in the public databases about their carcinogenicity in humans or animals are placed in IARC groups 1, 2A and 2B. For

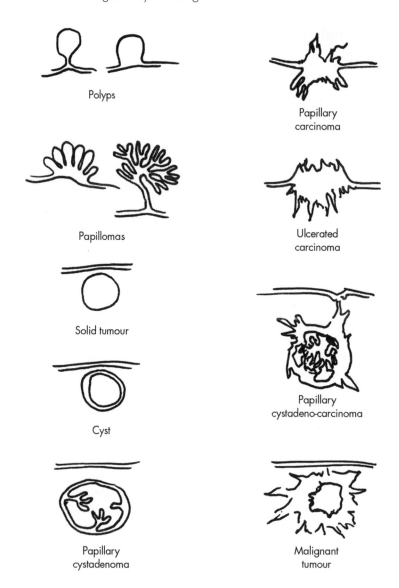

Polyps

Papillary carcinoma

Papillomas

Ulcerated carcinoma

Solid tumour

Cyst

Papillary cystadeno-carcinoma

Papillary cystadenoma

Malignant tumour

Figure 6.1 Various types of benign and malignant neoplasms.

many drugs such data are only confidentially available within the company, so that such drugs cannot be classified and end up in IARC group 3: not classifiable as to its carcinogenicity to humans. The IARC group 4 contains agents that are probably not carcinogenic to humans; so far caprolactam is the only chemical in this group because it is very hard to prove unambiguously that a chemical is not carcinogenic to humans. That requires extensive human exposure, including good epidemiology!

Mechanisms of carcinogenicity: initiation, promotion and progression

Changes in gene expression are required to change a cell from its normal, regulated cell division pattern to the uncontrolled situation seen in a neoplasm. A major requirement is that a cell becomes mutated in critical genes controlling cell division. Once such mutations have occurred (in principle, several mutation are needed to reach that stage), the cell can divide to form a clump of cells that sooner or later may cause local malfunction. If it becomes malignant it may cause such malfunction at sites of metastases. Figure 6.2. shows how a number of mutations have been identified as critical in the steps from a normal colorectal epithelial cell to a metastatic carcinoma that may metastasise to the liver for instance. The requirement for several mutations may be a reason why it takes (many) years for an initiated cell to grow out to a clinically relevant tumour ('cancer').

Chemicals that cause such mutations are called 'genotoxic', as discussed in Chapter 5. These chemicals can convert a normal cell to an 'initiated cell', a cell that has the potential to grow out into a tumour. However, in order to do so it requires more mutations and requires growth stimulation ('promotion'). Thus a combination of mutations and growth stimulation results in progression from the initiated cell to the disease 'cancer'. This model (initiation, promotion, progression) has some important implications for safety assessment of drugs. If a chemical increases the number of neoplasms in animal experiments, it

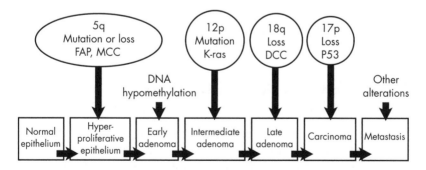

Figure 6.2 Pathogenesis of human colorectal cancer. The figure indicates mutations required to generate a malignant colorectal tumour type.

is very important to know whether it does so by a genotoxic event (and, thus, *causes* cancer) or by a growth-stimulating, 'epigenetic' effect (and, thus, does not cause cancer by itself but *promotes the outgrowth* of an existing initiated cell).

'Epigenetic carcinogens' are chemicals that increase the incidence of tumours by a non-genotoxic mechanism ('promotion') and thus require the presence of initiated cells; they can do this by several mechanisms, such as increased cell proliferation in response to cytotoxicity, hormonal stimulation, or inhibition of the immune system. Promoters stimulate the outgrowth of an as yet 'silent' initiated cell that, without such a growth stimulus, would not have had a chance to grow out to a tumour but might have been killed by the immune system ('immunosurveillance'). Inhibition of apoptosis (which will normally occur if cells lose their contacts on their way to metastasis) and stimulation of angiogenesis, required for the growing tumour, are important for progression of a tumour.

On the basis of this somewhat simplified initiation/promotion model, it is essential to establish whether a compound has genotoxic properties (Chapter 5): if it does, it may be an 'initiator' and pose a serious risk of *causing* cancer. The important practical issue is that for genotoxic carcinogens no safe threshold exposure is assumed to exist: theoretically, a single molecule could cause the critical mutation in a cell.

For promoters, however, a safe exposure level (i.e. dose) may be determined. For instance, if a compound promotes outgrowth of an initiated breast tumour cell owing to its oestrogenic properties, it will not have an effect if the dose is below the minimum required for its oestrogenic effect. If the dose required for its therapeutic effect is sufficiently below this level, it may be safe to use as long as the margin of safety (see below) is adequate.

For drugs, the implication is that genotoxic chemicals will not be further developed unless the intended indication allows this, such as cancer or any other life-threatening disease.

When is a carcinogenicity study required for a medicinal product?

If the exposure to a (non-genotoxic) chemical or drug is very short (e.g. a few days or weeks at low level) it is supposed not to cause cancer; for non-genotoxic drugs this seems a reasonable assumption. Only for drugs that are indicated for long-term treatment is a carcinogenicity test in animals required. Using European guidelines this is the case when:

- The drug is likely to be administered regularly over a substantial period of life (continuously during a minimum period of six months, or frequently in an intermittent manner so that the total exposure is similar);

or

- Where a substance has a chemical structure that suggests a carcinogenic potential;

or

- Where a substance causes concern due to:
  - Some specific aspects of its biological activity (e.g. a therapeutic class of which several members have produced positive carcinogenic results)
  - Its pattern of toxicity or long-term retention (of substances or metabolites) detected in previous studies
  - Findings in mutagenicity tests and/or short-term carcinogenicity tests

If the indication is for certain serious, life-threatening diseases such as certain forms of cancer, the requirement for carcinogenicity testing can be waived. Quite a few cytostatic drugs are carcinogenic *because of* their therapeutic mechanisms of action (e.g. they kill a tumour cell by causing DNA damage!), in which case a carcinogenic effect is unavoidable. Because the life expectancy for some diseases is very short, an eventual carcinogenic effect after several years would be offset by an increased lifespan (with sufficient quality of life). For instance, when the first drugs against AIDS were developed, they were allowed onto the market without having been tested for carcinogenicity. Under pressure from patients, who accepted the cancer risk rather than imminent death, the requirement was waived. Now, though, with the availability of effective anti-AIDS, drugs the requirement for proven non-carcinogenicity of new anti-AIDS drugs is expected to be enforced again.

Assessment of the potential carcinogenicity of drugs

For the reasons mentioned above, most candidate drugs that have to be tested for carcinogenicity will be devoid of genotoxicity, since the company selects only those for further development. Therefore, if the compound turns out to be carcinogenic in animals, it will most likely be due to a non-genotoxic effect. The mechanism of its non-genotoxic carcinogenic action then needs to be defined in order to assess the margin of safety of this drug during clinical practice.

The *in vivo* carcinogenicity assay

Testing for carcinogenic potential is usually done in rats and mice; occasionally hamsters are used. The IARC has decided that in the absence of adequate data on humans, it is biologically plausible and prudent to regard agents for which there is sufficient evidence of carcinogenicity in experimental animals as if they represent a carcinogenic risk to humans. Nevertheless, in its classification of compounds (Table 6.1) it does not classify a compound in group 1 unless there is factual proof that it is carcinogenic in humans. Obviously, for many chemicals such proof will not be available because exposure is too limited or the compound has never been investigated epidemiologically. The same problem is encountered for medicinal agents: unless a specific epidemiological survey is done, a 'mild' (i.e. low-incidence) carcinogen may go undetected. For instance, several local anaesthetics are genotoxic. Because they are used only occasionally this does not cause much concern, but it cannot be excluded that they might have a carcinogenic effect. Since such an effect is not apparent, the conclusion is that *if* they are carcinogenic then the incidence probably is very low.

A complete carcinogenesis experiment in rats requires administration of the drug to a sufficient number of animals (usually 50 per group per dose per sex) for most of their lifetime (i.e. usually 2 years), starting after they reach young adulthood. The route of administration should be the same as the therapeutic route. It is obvious that inhalatory exposure will lead to a very different exposure pattern (in particular in the nasal cavity, throat and lung) from that due to oral exposure to the same compound. Usually the route is oral by gavage or mixed with the animal food, but dermal or inhalatory exposure can be used. At least three dose levels are used, of which the highest traditionally is so high that it has some toxic effect on the animals (the maximum tolerated dose (MTD)). For instance, the increase in body weight is reduced in comparison with the control animals. The effect should not be such that the treated animals die earlier than the controls. The second dose could be one half the MTD, while the third would be again one half or one quarter of this middle dose. Using a satellite group of rats, the toxicokinetics of the drug will be followed over time, so that the internal exposure to the agent can be compared to that in the patients on a therapeutic dose schedule; this allows determination of a margin of safety.

At the end of the treatment period the rats are killed and pathologists will evaluate the tissues for the presence of tumours. The assignment of tumour type (benign, malignant) is critical; tissue slices and blocks will be stored for later re-evaluation in case of ambiguity. If an increased tumour incidence is observed, it will be analysed statistically for a trend. There are many cases where an increase in tumour incidence is observed for a non-genotoxic carcinogen. In that case the company will have to explain why this effect is irrelevant for the human patient (if in fact it is). For example, a drug may increase the incidence of thyroid tumours in the rat because it induces the glucuronidation of the thyroid hormone thyroxine, and thereby increases its rate of elimination. As a result, the thyroid-stimulating hormone (TSH) rises, resulting in increased cell turnover in the thyroid. In humans, thyroxine is mainly sulfated and consequently such an increase in TSH upon treatment with the same drug is not observed. For that reason it is accepted that such drugs are not likely to cause thyroid tumours in human patients and the animal finding is disregarded.

Tamoxifen carcinogenicity

An example of the outcome of a carcinogenicity experiment is shown in Table 6.2. The anti-oestrogenic drug tamoxifen was administered for a period of 2 years orally at three dose levels in male and female rats. A dose-related increase was observed in both benign and malignant hepatocellular tumours in both sexes, and in metastatic tumours in males. On the other hand, (natural) breast cancer in females was virtually completely prevented. This nicely illustrates how difficult it may be to judge the safety/efficacy balance: clinical data also showed that tamoxifen prevents oestrogen-dependent breast tumours in women, but how much risk of liver tumours do they then run?

Species differences and the *in vivo* assay

The mouse is often used, but results are considered less useful than those for rats. A major reason is that the spontaneous tumour incidence in many mouse strains is much higher than in the rat in a 2-year study. Thus, their power is less than that of the rat. It is of interest to note that there may be pronounced differences in tumour sites and frequencies between mouse strains. Attempts are under way to use transgenic mice,

Table 6.2 Carcinogenic effects of tamoxifen in the rat

Site/effect	Male Number of rats				Female Number of rats			
	102	51	51	50	104	52	52	52
	Dose (mg/kg/day)				Dose (mg/kg/day)			
	0	5	20	35	0	5	20	35
Hepatocellular								
Adenoma (benign)	1	8	11	8	1	2	6	9
Carcinoma (malignant)	1	8	34	34	0	6	37	37
Metastatic to lung or lymph node	0	0	3	4	1	0	2	2
Mammary gland								
Fibroadenoma (benign)	1	0	0	0	16	0	1	0
Adenocarcinoma (malignant)	0	0	0	0	9	0	0	0

The tamoxifen was administered daily by gastric intubation as a suspension. Data taken from Greaves P, Goonetilleke R, Nunn G, Topham J, Orton T (1993). Two-year carcinogenicity study of tamoxifen in Alderley Park Wistar-derived rats. *Cancer Res* 53(17): 3919–3924.

which might show a response to carcinogens within a shorter time (e.g. 6 months instead of 2 years).

There are many compounds that are carcinogenic in only one of the two species, or in only one organ. If a compound is carcinogenic in both species tested, at several sites, it usually is considered very likely that it is also carcinogenic in humans, even if it appears non-genotoxic.

As outlined above, the animal carcinogenicity assay lasts only 2 years: the lifespan of a rat or a mouse. In humans, however, it usually takes longer for cancer to develop: assuming that in many cases several mutations are required, and that cell growth from the single-cell stage to the detectable tumour requires many rounds of cell division, it appears that usually some 20 years of more or less continuous exposure is required. It is implicitly assumed in animal carcinogenesis experiments in rat or mouse that this 2-year exposure experiment can be compared to the human lifespan of 70–80 years. This is why a 2-year exposure in the rat is considered sufficient for humans, while a 2-year exposure in cynomolgus monkeys (which live much longer) is insufficient. However, in some cases a tumour can be generated much more quickly from even single exposures. This has been used to establish an assay for carcinogenicity in which a newborn rat receives a single injection in the first few days after birth. Presumably owing to the rapid growth in the first few weeks of life, in this case the tumour develops much more rapidly, even after a single exposure to a genotoxic compound. Interestingly, the inadvertent exposure of people to aristolochic acid, which is present in certain herbal preparations, caused urothelial malignancies within a relatively short period after (limited) exposure, presumably because of its strong genotoxicity (initiation) in combination with high nephrotoxicity (promotion).

Summary

The broadest definition used for a chemical carcinogen is that it is an agent that will increase the number of tumours, and/or decrease the time required for tumours to appear, in experimental animals and/or in humans. Such a broad definition does not consider many important aspects that must be addressed for the safety evaluation of a given carcinogen: the species (humans or rats), mechanism of action (genotoxicity or promotion), type of tumour (malignant or benign) and the potency (single low dose or repeated exposure to very high doses).

The following situations can occur:

- The drug is genotoxic: it may be carcinogenic, which would exclude application for most diseases except rapidly life-threatening diseases for which there is no therapeutic alternative. Otherwise, only short-term use may sometimes be acceptable.
- The drug is not genotoxic but shows increased incidence of certain tumours. Then:
  - If an explanation can be provided why it cannot be carcinogenic in humans, there will be no restrictions for its use.
  - If the margin of safety is sufficiently large, its use in humans will be acceptable.
  - If the drug is not genotoxic and not carcinogenic, no restrictions apply.

Protein-type drugs

A special case are the protein-type drugs, such as the (humanised) antibodies or human proteins (e.g. insulin analogues). Because these cannot be administered for a prolonged period to rats or mice, as a consequence of antibody formation in the recipient, it is impossible to evaluate their potential carcinogenicity by chronic, lifetime treatment. Their potential carcinogenicity will thus be difficult to assess. However, clinical signals of possibly increasing cancer incidence for some biologicals increase the urgency for appropriate assays.

It is reasonable to assume that these proteins do not cause mutations, yet a promoter effect remains a possibility. Whether the protein has growth-stimulating properties can be studied by effects on cell growth *in vitro*. However, if an antibody's therapeutic effect is to remove a human protein, for instance an antibody against tumour necrosis factor α, it is difficult to assess the effect of this treatment on the likelihood of a patient developing a tumour. Transgenic animals might be helpful in some cases, but will probably have to be constructed for that particular situation. It will become clearer with time whether there really is a problem with these proteins.

How can the carcinogenic action of a drug be discovered in humans?

An epidemiological study is required to find out whether administration of a drug causes cancer in patients. Unless the drug increases the incidence of very rare tumours or is very strongly carcinogenic, it might be quite difficult to pick up a carcinogenic effect. IARC has established that for a number of drugs the evidence in humans is sufficient to classify them in group 1 (Table 6.1). Thus, the carcinogenic action of *diethylstilbestrol* in humans was discovered because the vaginal tumours it induces are very rare in young girls (see Chapter 4). Similarly, the carcinogenic effect of *alkylating agents* was discovered because in young patients treated at high dose for Hodgkin's disease (a form of leukaemia), secondary tumours were observed some 20 years later.

Another class of human carcinogens are the *oestrogenic compounds*. In this case an increase of, for example, breast tumour incidence is observed, presumably as a consequence of oestrogenic stimulation. *Immunosuppressive compounds* like ciclosporin have been shown to be carcinogenic, presumably because they prevent the immune system removing initiated cells or small tumours.

If medicinal drugs are less strongly carcinogenic, it will be much harder to detect their effect. To achieve this, a large number of patients will have to be included with the appropriate controls, which is not easy and is very expensive; this is necessary because many confounders in lifestyle and disease have to be ruled out. Sometimes there are suggestions of an effect. For instance, diuretics may increase the incidence of renal cell carcinoma, or proton pump inhibitors may increase the number of gastric polyps or carcinoid tumours. However, in these and similar cases the evidence is often rather weak and the investigations required to prove or disprove these suggestions are not readily done. Thus, weakly carcinogenic medicines may escape detection.

The problems in epidemiological studies can be exemplified by phenacetin, an analgesic that has been off the market since the early 1980s because it was associated with kidney tumours. The epidemiological evidence was primarily based on groups of patients taking high doses of analgesics, including predominantly phenacetin, for prolonged periods. IARC therefore classified 'analgesic mixtures containing phenacetin' in group 1, while phenacetin as a single compound is classified in group 2A. It remains in 2A because there is not enough evidence in human patients for phenacetin *as a single compound* – the IARC classification is strictly evidence based!

Further reading

Franks LM, Teich NM (2001). *Introduction to the Cellular and Molecular Biology of Cancer*, 3rd edn. Oxford: Oxford University Press.

Klaasen CD, ed. (2001). *Casarett & Doull's Toxicology. The Basic Science of Poisons*, 6th edn. New York: McGraw-Hill, chapter 8.

7

Liver toxicity

J Fred Nagelkerke

Drug-induced hepatotoxicity is relatively common in patients. One reason is that its detection is very sensitive: even very mild liver toxicity is more easily detected (by clinical chemistry parameters in blood samples) than toxicity in most other organs. An even more important reason may be that after oral intake of drugs the liver is the most exposed organ in the body owing to first-pass uptake of drugs in that organ. In addition, the liver's capacity to metabolise drugs to reactive intermediates is very high. Finally, (carrier mediated) biliary excretion may result in high concentrations of drugs or their metabolites in bile; this may interfere with biliary excretion of other endogenous and exogenous substances, and may cause irritation of bile ducts.

Anatomy of the liver

The liver is a relatively large organ, comprising some 2% of body weight in adults and 4% in babies. In rats and mice it is some 4%. It receives blood (approximately 1500 mL/min) from two major sources. Arterial blood (300 mL/min) is supplied by the hepatic artery and is high in oxygen. Venous blood (1200 mL/min) is supplied from the intestinal tract via the portal vein; this blood is relatively low in oxygen but may be very rich in compounds absorbed from the intestine. Blood leaves the liver through the hepatic vein, containing compounds that were not taken up or were produced by the liver. The hepatocytes produce bile, which flows into bile canaliculi that come together in the bile duct (Figure 7.1).

The arterial and portal blood mix early in the liver unit, the sinusoid. If the liver sinusoid is seen as a 'tube' along which the hepatocytes are lined from portal vein to hepatic vein (Figure 7.2), the first cells are supplied with oxygen-rich, nutrient-rich blood ('zone 1') and the last cells of the sinusoid receive relatively low-oxygen blood ('zone 3'). These areas more or less coincide with the classical periportal and

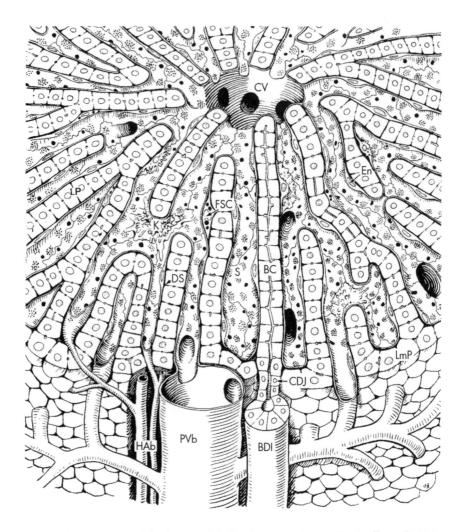

Figure 7.1 Diagram of the hepatic lobule. CV, central vein; K, Kupffer cell; FSC, fat storage cell; BC, bile canaliculus; En, endothelial cell; S, sinusoid; DS, Disse space; HAb, hepatic artery branch; PVb, portal vein branch; CDJ, caniculo-ductular junction; BDl, bile ductule; LmP, limiting plate; LP, liver plate. Modified from Motta P, Muto M, Fujita T (1978) *The Liver*, Tokyo, Igaku-Shoin.

centrilobular areas, respectively. This has toxicological relevance since the distribution of many enzymes is not homogenous across the liver: some are higher in zone 1, others in zone 3. For instance, several cytochrome P450s have higher activity in zone 3, while sulfotransferases are higher in zone 1. If a compound, such as paracetamol, is activated

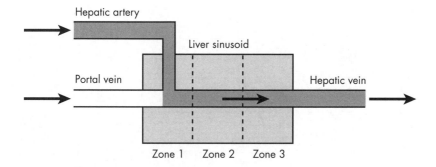

Figure 7.2 Schematic drawing of zones 1 to 3 in the liver, showing that arterial and portal blood mix in zone 1.

to a toxic intermediate by CYP activity, its toxicity is then observed in zone 3 rather than in zone 1. For compounds bioactivated by sulfation, the opposite applies.

The main liver cell type, the hepatocytes, are arranged in plates. Because of this architecture, all hepatocytes are polarised: the sinusoidal membrane faces the blood and the canalicular membrane faces the bile (Figure 7.1). This polarisation allows the hepatocytes to take up compounds from the blood (e.g. by carrier proteins) and secrete them (modified or unmodified) into the bile.

The lumen of the sinusoids is covered with endothelial cells. Unlike in other blood vessels, these contain large pores called fenestrae with an average size of 150–200 nm (Figure 7.3). Small compounds can therefore freely access the hepatocytes, but blood cells cannot.

The liver contains specialised macrophages, Kupffer cells. These are often present at branches of blood vessels and function in the removal of blood components (Figure 7.4). They secrete cytokines under various conditions, which may result in toxicity to hepatocytes. Pitt cells or hepatic natural killer (NK) cells have a somewhat obscure, probably immunomodulatory, function. Finally, stellate cells have been implicated in cirrhosis, which is a serious, irreversible, fibrotic liver disease resulting from long-term exposure to ethanol and some drugs.

Functions of the liver

Because all nutrients from food first enter the liver through the portal vein after their uptake from the intestines, the liver has an important role in their intermediary metabolism: fatty acids, carbohydrates and

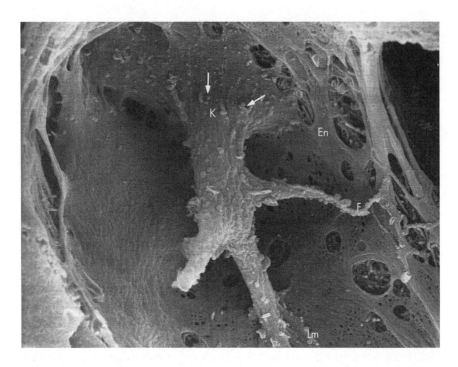

Figure 7.3 The Kupffer cells (K) are present in the sinusoid. They are in contact with the endothelial cells (En) through filipodia (F) and lamellipodia (Lm). Modified from Motta P, Muto M, Fujita T (1978) *The Liver*, Tokyo, Igaku-Shoin.

amino acids are taken up and extensively metabolised or incorporated into biomolecules in the liver. The hepatocytes closest to the portal vein (zone 1) in particular are exposed to high concentrations of nutrients. Several metabolic processes (e.g. glycolysis and gluconeogenesis) show functional differentiation between zone 1 and zone 3.

The liver plays a central role in intermediary metabolism of many essential endogenous compounds. For instance, the liver takes up excess glucose and stores it as glycogen in response to insulin-signalling when the glucose concentration in the blood increases. If blood glucose decreases, the liver releases glucose again under the influence of glucagon. It thus plays a role in glucose homeostasis.

Similarly, the liver plays a major role in the metabolism of fatty acids and cholesterol. Fatty acids can be broken down by β-oxidation or transported in the form of lipoproteins. Various kinds of lipoproteins are taken up by the liver and processed or stored for subsequent metabolism. Cholesterol is converted to bile acids, which are essential components of bile.

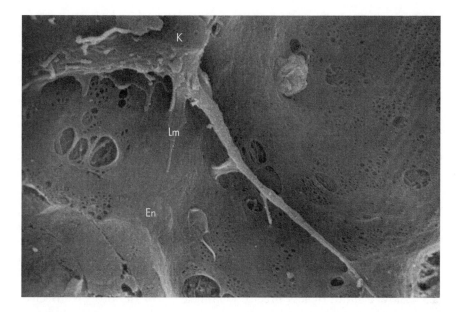

Figure 7.4 Interaction between a Kupffer cell (K) and endothelial cells (En). The endothelial cells contain many larger and smaller fenestrations. The lamellipodia (Lm) have grown through the fenestrations and are apparently in contact with the underlying hepatocyte. Modified from Motta P, Muto M, Fujita T (1978) *The Liver*, Tokyo, Igaku-Shoin.

The intermediary metabolism of amino acids takes place in the liver. Their degradative metabolism leads to the production of ammonia, which is subsequently rapidly converted to urea in that form excreted via the kidneys. A severe loss of liver function often leads to a rise of ammonia in blood, which is very toxic to brain function.

The liver synthesises many essential blood proteins, for instance albumin, clotting factors such as thrombin, the von Willebrand factor, factor XII, as well as lipo-apoproteins and other transport proteins.

The above functions are typical hepatocyte functions, while the other cell types each have their own critical roles to play. For instance, the Kupffer cell is essential for the elimination of particulate matter from the blood and subsequent degradation.

Of diagnostic importance is the glucuronidation of bilirubin, the breakdown product of the haem group: hepatotoxicity often results in impaired metabolism of bilirubin, leading to jaundice. This illustrates the important role of the liver in the metabolism of many endogenous components such as thyroxine and steroid hormones to breakdown products or metabolites to be excreted in bile or urine.

Quantitatively, most of the biotransformation of xenobiotics takes place in the liver. The metabolites may be released into bile or into the blood to be excreted in urine (the route for most low-molecular-weight substances). In this way the liver plays an extremely important role in the elimination of drug metabolites from the body. A decreased liver function, therefore, may be a contraindication for drugs that are mainly metabolised in the liver and have a small margin of safety.

Hepatotoxic compounds can affect the above processes in a highly specific way, or they may have a general effect on many of them at the same time. As a specific effect, for example, a compound may selectively inhibit the glucuronidation of bilirubin, leading to jaundice of the unconjugated type (i.e. a high unconjugated bilirubin concentration in blood). Alternatively, inhibition of the biliary excretion of the bilirubin conjugate in bile may lead to jaundice of the conjugated type (i.e. high concentrations of bilirubin glucuronide in blood). Such specific changes may allow a more or less precise identification of the process affected by the toxicant.

Types of liver toxicity

Several major types of liver toxicity can be distinguished, each characterised by typical clinical and diagnostic features. Some toxicity is irreversible, such as cirrhosis or liver cancer. Other effects can be overcome relatively easily because the liver has a tremendous capacity for regeneration. For diagnostic purposes it may be necessary to take a liver biopsy with a fine needle in order to assess the nature of the clinically observed loss of liver function. Histological evaluation of tissue slices, aided by special staining techniques, will allow conclusions that cannot be reached from blood biomarkers alone.

Liver necrosis

When drugs give rise to severe cytotoxicity in the liver, for instance a high acute dose of paracetamol (12 grams and more), massive necrotic cell death will occur. Hepatocytes disintegrate and a characteristic infiltration of lymphocytes occurs. If in the course of this process more than 80% of the hepatocytes are rapidly lost, the damage may become irreversible and lead to complete loss of liver function. Usually the patient dies after one or two days owing to a steady rise in ammonia in the blood, which severely damages brain function. If the damage to the liver is more limited, the patient may recover when the liver gets a chance to regenerate by rapid cell division. It may then grow out to the original size after a few weeks.

Cirrhosis

When the cytotoxicity is a continuing process, for instance due to chronic excessive alcohol intake, regeneration of hepatocytes may not take place, and lost hepatocytes may be replaced by fibrotic tissue. Hepatic fibrosis is a result of an imbalance in the production and degradation of extracellular matrix (ECM). The ECM is composed of collagens (especially types I, III and V), glycoproteins and proteoglycans. Stellate cells, located in the perisinusoidal space, produce ECM. After damage to the liver, cytokines (e.g. transforming growth factor β1) are secreted, which stimulate the stellate cells to produce ECM. Increased ECM deposition in the space of Disse (the space between hepatocytes and sinusoids) and a reduction in the size of the fenestrae of the endothelial cells contribute to the perturbation of the blood flow in the liver. Both the formation of plates of fibroblasts and the deposition of ECM lead to the development of portal hypertension. This is a progressive process that ultimately leads to replacement of all functional liver tissue by non-functional connective tissue. Initially this does not affect the person because of the overcapacity of the liver. At a certain stage the damage becomes apparent. A common sign is increased blood pressure because of the resistance of the cirrhotic liver to blood flow. As a result, the blood by-passes the liver owing to the formation of collateral blood vessels and liver function may cease totally, leading to death. This cirrhotic process is irreversible.

Steatosis ('fatty liver')

Several agents, including ethanol, tetracyclines and valproic acid, disturb the uptake and export of lipids and fatty acids into and from the liver. This may lead to excessive storage of lipids in the hepatocytes. The amount of fatty acid in the liver depends on the balance between the processes of delivery to the liver by chylomicrons and removal by very low-density lipoprotein (VLDL). Remarkably, pregnant women also develop steatosis.

When ethanol consumption is stopped, fat storage returns to basal levels within 2–4 weeks. Histologically, fatty liver is characterised by fat accumulation, which is most prominent in the centrilobular zone 3. The hepatocytes each contain one or more large (sometimes huge) fat droplets that displace the nucleus to an eccentric position. Early changes observed by electron microscopy include accumulation of membrane-bound fat droplets, proliferation of smooth endoplasmic reticulum, and gradual distortion of mitochondria.

By itself, steatosis is not dangerous and is reversible: the cells continue to function. When the inducing factor is removed, the cells regain their normal appearance and functionality. However, the Reye-like syndrome, characterised by acute microvesicular steatosis and hyper-ammonaemia, in infants and young children upon salicylate treatment may be lethal.

Cholestasis

The formation of bile (requiring excretion of bile acids into the bile canaliculi) is an active process in which many cellular proteins are involved, including transporters for bile acids. Some drugs interfere with these regulatory proteins by competition with their endogenous substrates or by direct inhibition, resulting in cessation of bile flow or selective inhibition of the biliary excretion of specific compounds. In that case products are no longer secreted from the liver into the bile, result-ing in their accumulation in the body. A well-known example is the retention of the haem breakdown product bilirubin. Normally this leaves the body via biliary secretion. If this process is blocked, bilirubin accumulates in skin and the white of the eyes, which then turn yellow (jaundice). Cholestasis, including accumulation of bile salts in blood because of decreased bile production (due, for example, to loss of carrier-mediated biliary excretion of these components), may be observed during therapy with phenothiazines, erythromycin, sulfon-amides or oral contraceptives.

Hepatitis

Some compounds can change hepatocytes in such a way that cells from the immune system receive a signal to remove these liver cells. A well-known example is hepatitis after halothane exposure. It appears that halothane is converted into a reactive intermediate that binds to endogen-ous proteins. Halothane thus provides a hapten, a chemical group that is recognised by the immune system as 'foreign'. This stimulates an attack of immune cells on these hepatocytes, leading to necrosis and apoptosis. Histologically this is characterised by an infiltration of immune cells into the liver as if there were inflammation. Many drugs cause hepatitis, usually with low frequencies, such as isoniazid, tricyclic antidepressants, phenytoin and sulfonamides.

Liver cancer

Several compounds cause cancer in the liver in animal experiments. As discussed in Chapter 6, this may be due to genotoxic or non-genotoxic ('epigenetic') mechanisms. In the latter case a sufficient margin of safety may allow the therapeutic use of the drug. In any case the occurrence of liver carcinomas in long-term animal assays will have to be explained before such a drug can be registered, because it is an irreversible effect. Some anabolic steroids may cause hepatocellular carcinomas.

Consequences of liver failure

In view of the central role of the liver as described above, it is clear that serious liver damage has severe implications for the patient. For instance, if glucose homeostasis is lost, the brain cannot function properly because it depends on a constant supply of glucose. The most prominent result of liver failure is the loss of processing of ammonia into urea: if this process fails, blood ammonia will increase, resulting in poisoning of the brain: 'hepatic encephalopathy' will lead to death.

Assessment of liver toxicity: biomarkers and diagnostics

Several diagnostic procedures are available to assess loss of liver function by toxicity. Clinically, jaundice is a sign of serious liver toxicity, both in patients and in animals: blood levels of (unconjugated) bilirubin increase so much that in particular the white of the eyes is visibly yellow. Detection is possible by measuring the activity of certain enzymes, the aminotransferases or 'transaminases', in serum: alanine aminotransferase (ALT) and aspartate aminotransferase (AST) give a good indication of the extent of liver damage. Increases of these enzymes or of bilirubin in blood are routinely monitored in animal experiments and clinical trials to detect hepatotoxicity at an early stage. Increases of the transaminases three-fold or four-fold above 'normal' values are indicative of toxicity, especially if bilirubin is also increased markedly. For assessment of changes in liver function in more detail, examination of liver biopsies by histological means is useful. It is not uncommon that a mild increase in transaminases is transient: they may decrease again upon continuation of the therapy.

Unexpected liver toxicity may occur at very low frequency in patients, without any 'warning' from animal experiments or the clinical trials: if the incidence is 1:1000 patients or less, it will not necessarily be detected in regular clinical trials during Phase III. Unfortunately, such idiosyncratic reactions will only be picked up after market introduction of the drug. They may be of an immunological type, as can be confirmed by rechallenge of the patient.

Some examples of hepatotoxic drugs

Paracetamol (acetaminophen)

The analgesic paracetamol (acetaminophen in the USA) can cause hepatic necrosis when taken in overdose. Standard analgesic doses of paracetamol (1–3 g a day) are conjugated in the liver with sulfate or glucuronic acid and are excreted in urine. They are safe for almost every patient. However, in overdose (10 g and more) paracetamol is increasingly activated by cytochrome P450 isozymes to the reactive N-acetyl-p-quinone imine (NAPQI; see Chapter 2, p. 62). This binds very rapidly to the thiol group of the tripeptide glutathione (GSH), forming a non-toxic conjugate. However, this process of detoxification depletes GSH at high paracetamol doses. When GSH is no longer available, the imido-quinone binds covalently to cysteine-thiol groups on proteins, forming covalent paracetamol–protein adducts. N-Acetylcysteine, which restores GSH, prevents toxicity if it is given on time after paracetamol overdose; this compound is a standard medication after overdose of paracetamol. Using radiolabelled paracetamol and subsequent gel electrophoresis, it was shown to bind to several cellular proteins and the extent of this binding correlated with hepatic toxicity. The toxicity is seen particularly in zone 3 (the centrilobular areas of the liver) because that is where the bioactivating CYPs are located. The cells die by necrosis, which leads to infiltration of lymphocytes in the liver to remove the cellular debris. As long as the damage is limited and enough of the hepatocytes are left over (some 20% of the liver), the liver may regenerate and replace the lost cells. However, if the damage is beyond that limit, the loss of liver function is irreversible and the patient will die after one or two days because of fulminant hepatitis with massive cell death and infiltration of lymphocytes.

Damage to mitochondrial proteins and proteins involved in cellular ion control, as well as oxidative stress, underlie the liver toxicity of paracetamol.

Halothane

Halothane and other halogenated anaesthetic agents such as isoflurane may cause rare but severe liver dysfunction. Approximately 20% of halothane is oxidatively metabolised compared to only 0.2% of isoflurane. Toxicity of the anaesthetics is due to oxidative metabolism and subsequent binding of a reactive intermediate to liver proteins. The intermediate induces trifluoroacetylation of proteins, and this hapten on some proteins then becomes antigenic (see Chapter 2, pp. 63–64 and Chapter 10), leading to the formation of antibodies in sensitive individuals. Re-exposure to halothane initiates an attack of immune cells on the liver; this is how the immune character of halothane toxicity in humans was discovered. The patient develops a fever, and in a liver biopsy sample the presence of leucocytes and other eosinophilic cells is histologically evident. The incidence of this type of liver toxicity is relatively rare: only 1 case per 6000–35 000 patients after exposure to halothane. Fulminant necrosis after anaesthesia with such halogenated agents therefore occurs only in susceptible individuals.

Isoniazid

Isoniazid may cause hepatitis-like liver toxicity after a few months' treatment: necrosis is observed in the liver. In many patients only a transient increase in transaminases is observed, but in some it leads to a severe reaction. Both genetic and environmental factors affect the reaction. In young patients it occurs rarely, while in patients aged 50–65 years it is more common. Acetylation is the major metabolic route of isoniazid, but its role in this toxicity is still unclear. However, interaction with rifampicin increases the hepatotoxicity of isoniazid strongly, suggesting that induction of metabolism of isoniazid to a toxic metabolite may play a role.

Tricyclic and tetracyclic antidepressants

Several drugs of this class, such as amitriptyline, imipramine, desipramine and amineptine, cause minor degrees of liver damage as shown by an increase of transaminases or (conjugated) bilirubin. In some cases, severe cholestasis may occur. High fever and eosinophilia may accompany these manifestations of liver toxicity. While in most patients these signs are mild and transient, in a small percentage of patients they may be severe.

Oral contraceptives

Ethinylestradiol and similar steroids may cause jaundice due to their cholestatic effect. Incidences were initially relatively high but rapidly decreased once low-dose formulations came onto the market. Signs of liver toxicity such as jaundice, pruritus and anorexia became apparent within a few weeks at the high doses. The oestrogens interfere with bile formation and biliary excretion of bilirubin and bile salts. The incidence of cholestasis is quite low at the much lower doses currently used.

Further reading

Klaasen CD, ed. (2001). *Casarett & Doull's Toxicology. The Basic Science of Poisons*, 6th edn. New York: McGraw-Hill, chapter 13.
Zakim D, Boyer T (2002). *Hepatology. A Textbook of Liver Disease*, 4th edn. Phildelphia: Saunders, section IV.

8

Kidney toxicity

Bob van de Water

The kidney has many essential transport functions to preserve body homeostasis. These processes require a high energy level, so that the organ is very sensitive to loss of ATP. Because xenobiotics and their metabolites are very often actively excreted and accumulated by kidney cells, these may become extensively exposed; in particular, cells lining the proximal tubule are the target for a wide variety of compounds. It is therefore not surprising that nephrotoxicity is a clinical problem for various classes of drugs.

Renal structure and function

The major functions of the kidney are related to its role in maintaining the homeostasis of the internal environment: the kidney excretes waste products and regulates the volume of the internal environment as well as the electrolyte and acid–base balance. To do this the two kidneys receive approximately 25% of the cardiac output: 1.2–1.4 litres of blood per minute in an adult person. The kidneys are also essential in the production of several important regulatory factors such as renin, erythropoietin and 1,25-dihydroxyvitamin D_3.

The kidney has several subregions: the cortex (cortical labyrinth and medullary rays), the outer medulla (outer stripe and inner stripe), and the inner medulla (the tip and the papilla). The smallest functional subunit in the kidney is the nephron (Figure 8.1), of which the kidneys contain 10^4 to 10^6 depending on the species. Some nephrons are superficial, and do not extend all the way into the medulla; deeper nephrons have large extensions into the medulla.

A nephron consists of distinct parts: the glomerulus, the proximal tubule (convoluted and straight tubule), the loop of Henle, the distal convoluted tubule, and the collecting duct (cortical, outer medullary and inner medullary collecting duct). The nephron stretches from the outer cortex (glomerulus) deep into the medulla (intermediate tubule) back to

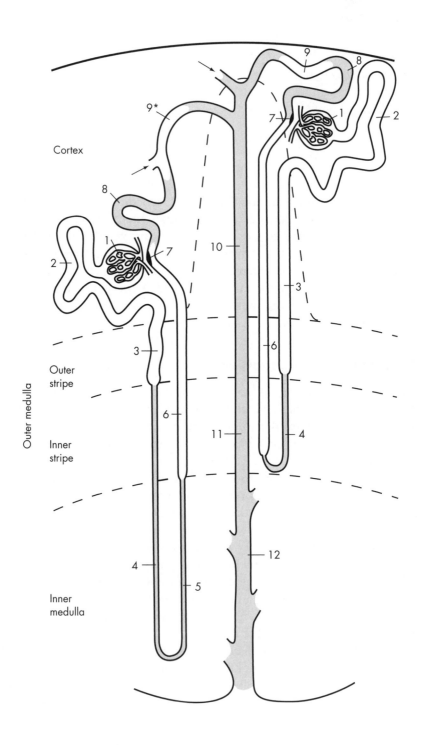

the cortex (distal tubule and collecting duct). It plays a pivotal role in kidney functions: each specific part of the nephron has its specialised task. Thus, some 15–20% of the incoming renal plasma flow (approximately 700 mL plasma/min) is filtered by the glomeruli to produce an ultrafiltrate (125 mL/min). This ultrafiltrate contains the blood components of molecular weight below roughly 50 kDa, or size smaller than 75–100 Å. The proximal tubule reabsorbs about 50–60% of the ultrafiltrate. In this part of the nephron waste products are excreted and essential nutrients are reabsorbed by several active transport carrier systems. The balance of cations (sodium, potassium, magnesium, calcium), anions (chloride, bicarbonate, phosphate) and water is also controlled. The intermediate tubule is also involved in concentrating the urine: 20–30% of the filtered sodium and potassium, and 15–20% of the filtered water are reabsorbed. In the distal tubule most of the remaining water and sodium are removed, whereas potassium is either secreted or reabsorbed depending on the need. Finally, the collecting duct also reabsorbs water and sodium, but its contribution is quite small compared to that of the distal tubule.

Conditions that determine kidney toxicity

Over 300 chemicals have been identified to cause renal damage. Many are still used in clinical practice or industry, which are the major sources of human exposure to nephrotoxicants. Although various segments of the nephron can be a target for nephrotoxicants, the proximal tubule is the most common site for chemically induced nephrotoxicants, including drugs (Table 8.1): antibiotics such as cephalosporins and aminoglycosides, radiocontrast agents, chemotherapeutics, immuno-suppressives and analgesics. Among industrial and environmental pollutants, heavy metals including cadmium and mercuric chloride are prime examples. The proximal tubular cells are the primary target of nephrotoxicity by drugs. This is due to (1) the large volumes of fluids

Figure 8.1 A long-looped (left) and a short-looped (right) nephron together with the collecting system. Different segments of the nephron are indicated by numbers. (1) Bowman's capsule and glomerulus; (2) proximal convoluted tubule; (3) proximal straight tubule; (4) thin descending limb; (5) thin ascending limb; (6) medullary thick ascending limb; (7) macula densa; (8) distal convoluted tubule; (9) connecting tubule; (10) cortical collecting duct; (11) outer medullary collecting duct; (12) inner medullary collecting duct. Modified from The Renal Commission of the International Union of Physiological Sciences (IUPS) (1988) *Am J Physiol* 254(1 Pt 2): F1–F8.

Table 8.1 Medicines with nephrotoxic side-effects

Class	Specific examples
Antineoplastics	Cisplatin
	Doxorubicin (adriamycin)
	Methotrexate
Anti-infectives	Aminoglycosides
	Beta-lactams
	Amphotericin
Analgesics	Paracetamol (acetominophen)
	NSAIDs
Immunosuppresants	Ciclosporin
	FK-506
Radioconstrast agents	

NSAIDs, nonsteroidal anti-inflammatory drugs.

and amounts of blood-borne components (including toxicants) that are reabsorbed by and thereby accumulate in the proximal tubule, (2) the very high activity of many enzymes, and (3) the high metabolic rate, which is needed to keep all these processes going. The S_1, S_2 or S_3 segments of the proximal tubule are damaged, according to the physico-chemical properties of the therapeutic agents and the localisation of transport systems in the different segments. Such transporters include organic anion and cation carriers. The localisation of the carrier in the cell determines the potential concentration of chemicals at the apical membrane ('urine side') or basolateral membrane ('blood side') of the cell. On the one hand, accumulation of compounds in the proximal tubular cells occurs from the basolateral side through the action of, for example, organic cation and anion transporters. On the other hand, transport may take place from the apical side, where specific uptake occurs of cysteine conjugates through amino acid carriers or of low-molecular-weight proteins with the compounds bound to them, e.g. cadmium bound to metalloproteins.

The presence of enzymes that are involved in bioactivation or in detoxification of xenobiotics may determine whether a specific site of the nephron is affected. For example, the enzyme β-lyase metabolises cysteine conjugates of halogenated alkenes into reactive metabolites (Chapter 2). Although the latter pathway does not appear relevant for the nephrotoxicity of drugs, it stresses the potential involvement of 'non-classical' bioactivation pathways in relation to renal toxicity.

Types and consequences of renal cell injury

Accumulation of compounds and/or their bioactivation results in the injury of cells. Bioactivation generally results in the formation of reactive (electrophilic) intermediates that will affect the (macro)molecules in cells, e.g. glutathione, DNA or proteins. Alternatively, compounds themselves may have properties that directly affect biological processes by selectively inhibiting or activating enzyme activity or causing the increased formation of reactive oxygen species. As a result, detoxifying and/or repair processes will be initiated in the cell. For example, reactive oxygen species will be detoxified by antioxidant enzymes (see also Chapter 3). If the antioxidant pathways are insufficient, peroxidation of lipids occurs, affecting membrane integrity and permeability.

Besides membranes, the mitochondria are also an important cellular target for various nephrotoxic agents. ATP synthesis in proximal tubular cells is extremely critical because of their high energy requirement for active transport processes; damage to their mitochondria easily leads to ATP depletion. This will ultimately result in cell death. Severe cell injury due to high levels of reactive oxygen species and lipid peroxidation or massive ATP depletion results in necrotic cell death. When injury is more subtle but repair of the injury not possible, cells may initiate apoptotic cell death. Both types of cell death are observed after exposure to nephrotoxicants, but there is debate about the relative importance of each type (see Chapter 3).

Dead tubular cells are lost in the urine. When many cells die this can lead to obstruction of the tubule, resulting in an increased pressure upstream in the tubule that is compensated by vasoconstriction, resulting in decreased glomerular filtration. When this occurs in many nephrons, the overall glomerular filtration rate will drop. Another consequence of loss of tubular cells is the denudation of the basolateral membrane of the tubules. Since the tubular cells normally form the barrier between the luminal ultrafiltrate and blood, loss of the tubular cells results in a so-called back-leak of the ultrafiltrate from the lumen to blood compartment. This will hamper the clearance of components that should be removed from the body via the urine. When many proximal tubular cells are damaged at the same time, the overall function of the kidney will be affected within a couple of days, leading to acute renal failure.

Perturbations of renal function are not always acute. Severe DNA damage can result in the initiation of cell death, most often by apoptosis. However, when DNA damage is not so dramatic, but nevertheless inappropriately repaired, remaining DNA lesions may ultimately result in the development of renal cancer; this process takes many years to develop. Prolonged treatment with analgesics may not cause the direct injury of the kidney, but in the long term may initiate the infiltration of immune cells in the kidney. Although these immune cells may not cause direct acute nephritis, their continuous presence may gradually diminish renal function.

Clinical markers of renal failure

Renal clearance of compounds is a sensitive clinical and diagnostic parameter of renal function. For low-molecular-weight compounds that are *not protein-bound*, the following applies with normal kidney blood flow and kidney function. If a chemical is cleared by glomerular filtration only, its clearance will be ca. 125 mL plasma/min. If it is completely cleared by active transport at the basolateral side of the proximal tubule in the tubuli, its clearance will equal total renal plasma flow, i.e. 700 mL/min. For many compounds that are not actively excreted, the renal clearance is directly related to the glomerular filtration rate (GFR), i.e. the amount of fluid that is filtered by the glomeruli. Thus, lower GFR will lead to an accumulation of these compounds in blood. In the worst case, GFR drops to zero and these compounds can no longer be cleared from the blood (unless, for example, biliary excretion can take over). However, the kidney has a large overcapacity: it requires an 80% reduction of the number of functional nephrons (i.e. drop in GFR) before the overall kidney function is seriously impaired.

Several metabolic end-products are cleared by the kidney, including urea and creatinine. Therefore, measuring the levels of both creatinine and blood urea nitrogen (BUN) in blood gives an impression of renal function. Since BUN levels may also increase as a result of increased protein catabolism, creatinine is generally used as a more reliable marker for renal function. As well as creatinine levels in blood, the levels of creatinine in the urine are also required for accurate measurement of GFR. However, creatinine is cleared not only by glomerular filtration but also partially by tubular excretion. Therefore, for absolute values of the GFR, invasive studies are required that make use of the inert sugar inulin. Inulin is removed from the body solely by glomerular filtration. Thus, the exact GFR can be determined by measuring the levels of inulin in both the plasma and the urine.

The clearance of other compounds can be used to assess the function of tubular excretion. For instance, *para*-aminohippurate (PAH), which is injected intravenously for the test, is completely cleared by rapid tubular uptake from the blood compartment by an anion transporter followed by urinary excretion. Its rate of blood elimination reflects tubular function but, of course, is also a direct measure of the renal blood flow (which determines PAH supply to the tubuli).

Analysis of glucose concentrations in the urine gives information about tubular cell function: glucose should be reabsorbed at over 99%, so that glucose should be virtually absent from urine. Diabetes should obviously be excluded as the cause. Increased protein in the urine is also indicative of kidney problems. Large proteins such as albumin or immunoglobulin G are normally not filtered by the glomeruli and their presence indicates leakage of the glomerular filter, i.e. glomerular injury. When only small proteins that can pass the glomerular filter are observed in the urine, this could be indicative for proximal tubular injury.

Since the nephron consists of many functionally different cell types, cell-selective expression of specific enzymes will occur. Leakage of such enzymes from injured cells into the lumen and excretion into the urine could be a valuable cell-specific biomarker. Indeed, various markers have been found. For example, the presence of α-glutathione S-transferase in urine is indicative of proximal tubular injury. The presence of the membrane protein γ-glutamyl transpeptidase, which is present in the villi of proximal tubular cells, is also a good marker for proximal tubular cell injury.

The above are biomarkers for renal dysfunction. As discussed at the beginning of the section, these markers may only be picked up when relatively severe renal damage has occurred. Novel technologies such as metabolomics, which allows the simultaneous measurements of many different metabolites in the urine, may increase the sensitivity to allow accurate determination of renal dysfunction. Because of the analysis of many metabolites at the same time, a urine 'metabolite profile' may also allow the 'recognition' of the site where the injury took place in the kidney: e.g. glomeruli, proximal tubules or distal tubules.

Repair processes related to renal injury

Acute renal failure is associated with denudation and excretion of (often still viable) proximal tubular cells in the urine. This generally involves the cells from the S_3 segment of the proximal tubule. Upon severe injury, the loss of cells may extend all the way to the outer cortex and also affect the S_1 and S_2 segments of the proximal tubule, depending on the type and dose of a particular nephrotoxicant. The time course of the injury may also differ. For example, for ischaemia/reperfusion injury, the type of injury that is most important in the clinic, it takes up to one day to induce acute renal failure. In contrast, tubular damage to the proximal tubular cells by cisplatin and gentamicin generally takes several days to develop. The kidney has a high regeneration capacity, so that even when a large portion of the proximal tubular cells is lost, an outburst of proliferating cells is seen between day 1 and day 3 after the initial injury. These cells can be easily identified by protein markers that are indicative of cell proliferation and lack some properties of the normal proximal tubular cells: they are flattened and do not have large numbers of villi. This process is called de-differentiation. After 7 days, this proliferation is complete and is followed by a phase of re-differentiation. Cells will regain the presence of villi and other characteristics that are typical of proximal tubular cells. The injured renal tissue is repopulated with functional cells after approximately 3–4 weeks.

It is important to stress that the regeneration phase involves several processes: mitogenesis, migration of cells, morphogenesis and differentiation. Various growth factors and cytokines play essential roles in the regulation of the renal regeneration process, including adenine nucleotides, hepatocyte growth factor, epidermal growth factor, insulin growth factor, fibroblast growth factor and transforming growth factor, released from injured cells, from cells already present in the interstitium or from inflammatory cells that have infiltrated the tissue.

In vitro models for the study of nephrotoxicity

Renal (tubular) cells can be isolated from kidney (including human donor kidneys) and cultured *in vitro*. There are also available several renal epithelial cell lines that have characteristics comparable to the primary cultured renal cells. These primary cells and established cell lines have many of the characteristics that are relevant for nephrotoxicity, such as transport functions and enzyme activities. However, they have lost the *in vivo* environment, including the immune system, and have changed their metabolism. For example, *in vivo* the production of ATP in proximal tubular epithelial cells is largely mediated through oxidative phosphorylation in the mitochondria. *In vitro*, cells switch from oxidative phosphorylation to glycolysis. Renal cells *in vitro* are therefore less susceptible to inhibition of the activity of the mitochondria. In general, renal cell cultures are better models for investigation of molecular mechanisms of toxicity than studies with intact laboratory animals. However, the prediction of renal toxicity *in vivo* on the basis of *in vitro* experiments is difficult.

Examples of renal toxicity caused by therapeutics

Cisplatin

cis-Diamminedichloroplatinum (cisplatin) (Figure 8.2) is a potent chemotherapeutic agent that unfortunately has a narrow therapeutic window. Treatment of patients with cisplatin is limited by neurotoxicity, ototoxicity and nephrotoxicity. The proximal tubular cells are the primary target: cisplatin is taken up by the proximal tubular cells, most likely at the basolateral side. The highest concentration is reached in S_3, most probably related to the increased transport capacity, where injury is most severe. The concentration in the kidney can be decreased by addition of large volumes of saline. However, patients on cisplatin generally have a 10–30% loss of renal function.

The mechanism of cisplatin-induced renal dysfunction involves several steps (Figure 8.3). After cellular uptake, cisplatin loses two chloride ions and becomes hydrated. Hydrated cisplatin is less stable and can form double-strand cross-links with DNA. It can also react with RNA, phospholipids and amino acids present in enzymes and structural proteins. Accordingly, injury from cisplatin is also observed in the mitochondria, where the activity of enzymes of the respiratory chain is inhibited. These mitochondrial effects are most likely the cause of the observed increased oxidative stress in the cells. Oxidative stress in cells results in the activation of protein kinases that will affect gene expression and protein expression and thereby the overall biology of renal cells. Cisplatin also affects the organisation of the actin cytoskeleton network in cells; these changes have been implicated in the mechanism of the nephrotoxicity caused by cisplatin. The renal cell injury caused by cisplatin in cultured cells results in cell death, either by necrosis or by apoptosis.

Ciclosporin

Ciclosporin is a macrolide antibiotic composed of 11 amino acids. It is used as an immunosuppressive agent to allow both solid organ and bone

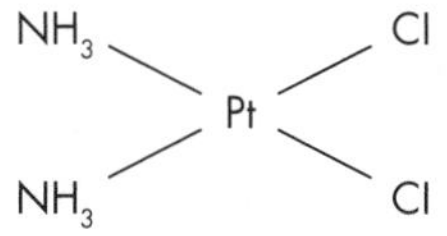

$$NH_3 \diagdown \diagup Cl$$
$$Pt$$
$$NH_3 \diagup \diagdown Cl$$

Figure 8.2 The structure of cisplatin.

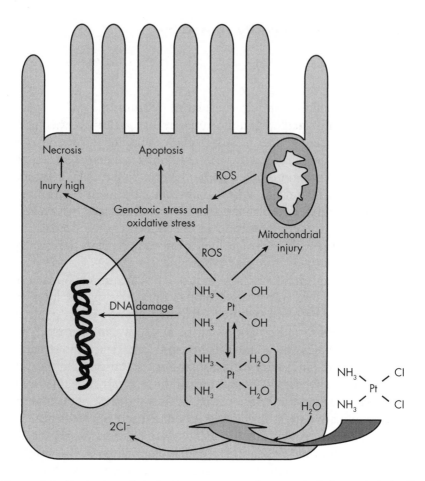

Figure 8.3 Mechanism of cisplatin toxicity in renal proximal tubular epithelial cells. (ROS, reactive oxygen species.)

marrow transplantation. Another immunosuppressant, FK506, with a similar mechanism of action, also has nephrotoxic properties. Toxicity of ciclosporin to the kidney remains an important problem. The exact mechanism of nephrotoxicity is not fully understood. There is evidence that ciclosporin has effects both on the proximal tubular cells and on the blood flow in the kidney. The effect on the proximal tubular cells *in vivo* is confined to the S_3 segment. This can be related to uptake of ciclosporin by these cells and/or additional effects of ciclosporin on the blood circulation in the kidney (see below). The toxic effect is associated with increased vacuolisation of the proximal tubular cells. The nephrotoxicity seems to be related to the induction of oxidative stress, which in part may be related to inhibition of mitochondrial function. Mitochondria in cells treated with ciclosporin are bigger and renal cells in cell culture change their metabolism from oxidative phosphorylation to glycolysis. But studies *in vitro* may not reflect the *in vivo* situation. In intact laboratory animals, ciclosporin also seems to affect the production of signalling lipids that directly affect vasoconstriction. These include thromboxanes and prostaglandins, which activate their receptors on blood vessels, thereby mediating vasoconstriction and limiting the blood supply to tubular segments. This may indirectly affect the viability of the proximal tubules in the S_3 segment. *In vivo*, both short-term (acute toxicity) and long-term (chronic toxicity) effects are observed. Long-term effects of ciclosporin are primarily interstitial renal fibrosis, as evidenced by regions in the cortex of the kidney that contain fibrotic tissue; though other areas may appear fully normal in the same kidney. It is unclear whether the chronic toxicity is based on the same effects as the acute toxic effects of ciclosporin.

Aminoglycoside and beta-lactam antibiotics

Antibiotics have long been a major cause of toxicity. Two major classes cause renal failure: aminoglycosides and lactam antibiotics. Most cases of toxicity are related to acute tubular necrosis resulting in acute renal failure. The kidney toxicity of these antibiotics lies in their route of elimination from the body being primarily through renal excretion.

Aminoglycosides are used to treat Gram-negative bacterial infection and tuberculosis. The natural compounds are streptomycin, kanamycin, tobramycin and the potent nephrotoxicant gentamicin. Typically approximately 10 times the therapeutic dose will result in acute tubular necrosis. Gentamicin has been studied most extensively since it is the aminoglycoside with the most prominent nephrotoxicity. The target of gentamicin is the proximal tubular cells. These cells take up gentamicin since it binds to phospholipids on the villi of proximal tubular cells. Next, gentamicin is taken up through endocytosis and ends up in the lysosomes. Since gentamicin binds to phospholipids, which then cannot be degraded, there is an accumulation of lipids and gentamicin in the lysosomes. This accumulation can be 10 to 100 times the concentration of that in the blood plasma. The consequence is that the lysosomes fail, but the link between lysosomal changes and cell injury is as yet unknown. Gentamicin also reduces mitochondrial respiration and protein synthesis.

Lactam antibiotics (penicillins, cephalosporins, carbapenems and

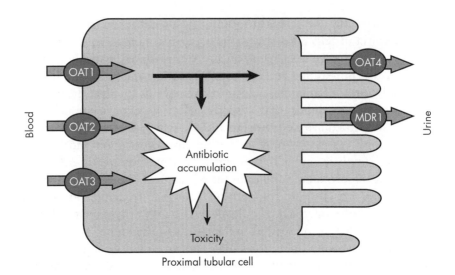

Figure 8.4 Mechanism of uptake of lactam antibiotics by organic anion carriers in proximal tubular cells. Lactam antibiotics are taken up by different organic anion transporters (OAT1, 2 and 3). This results in accumulation of the antibiotics in the proximal tubular cells, with potential cytotoxicity as a consequence. OAT4 and MDR1 are carriers that transport the substrates from the cell into the urine.

monobactams) are among the most important antimicrobials. Severe nephrotoxicity was rapidly discovered after the introduction of cefaloridine in the 1960s. This specific toxicity of cefaloridine was based on the specific structure (presence of a pyridinium ring), which is lacking in newer lactam structures which therefore have lower toxicity. Renal toxicity is in particular due to the active clearance of these compounds by the kidneys. This is related to the active uptake through organic-anion transporters present on the basolateral membrane of the proximal tubular cells (Figure 8.4). The specific uptake is the primary cause of targeting these cells. In the proximal tubular cells several biochemical perturbations take place, including oxidative stress and lipid peroxidation in the case of cefaloridine which was due to the pyridinium ring.

Later lactams do not cause oxidative stress. Lactam antibiotics also affect mitochondrial function.

Further reading

Tarloff JB, Lash LH (2005). *Toxicology of the Kidney*, 3rd edn. Boca Raton: CRC Press, 2005.

9

Toxicity in the respiratory system

Eva Brittebo

The respiratory system has direct contact with inhaled air that may contain a variety of environmental and occupational pollutants, noxious gases, dust, fibres and tobacco smoke, as well as airborne viruses, bacteria and fungi. This organ system may also be exposed to inhaled drugs used for local or systemic treatment of various diseases. Several drugs are administered via inhalators, currently mainly for local delivery. This route avoids first pass in the liver but is subject to first-pass uptake and metabolism in the conducting airway epithelium and lungs. Additionally, the lungs may receive a substantial amount of drugs and other foreign chemicals via the circulation because the lung receives the entire cardiac output.

Structure and function of the respiratory system

The inhaled air is moistened and warmed in the nasal cavity comprising a number of turbinates that increase the surface area. The nasal passages are covered with squamous, respiratory and olfactory epithelia in the different regions. Underneath the nasal epithelia there are numerous mucus-producing tubular glands and blood vessels.

The lung contains more than 40 distinct cell types. The trachea and bronchi/bronchioles are covered by ciliated cells and nonciliated cells such as bronchiolar Clara cells, goblet cells and basal cells, and supported by cartilage, connective tissue and smooth-muscle cells (Figure 9.1). Following damage to the airways, the Clara cells proliferate in order to regenerate the surface epithelium. The epithelium of the conducting airways is covered with a layer of mucus that is secreted by mucous glands, and contains secretory Clara cell proteins. The mucus is transferred towards the glottis by the action of cilia.

The principal function of the lung alveoli is gas exchange between air and pulmonary capillary blood. The flattened type I cells (pneumocytes) cover about 90% of the alveolar area and take part in the

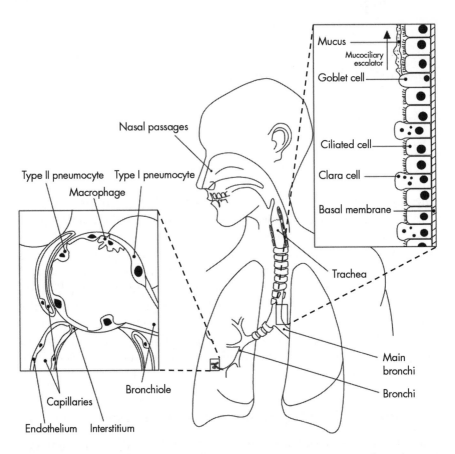

Figure 9.1 Schematic diagram of the human respiratory system. On the right is an enlarged section of the bronchial epithelium consisting of a number of ciliated and nonciliated cells covered with mucus. The columnar Clara cells have a high expression of drug-metabolising enzymes. On the left there is a cross-section of an alveolus with attenuated type I and cuboidal type II pneumocytes and an alveolar macrophage. The alveoli are in close contact with attenuated pulmonary capillary endothelial cells and these cells constitute the functional unit for gas-exchange. The small interstitium consists of fibroblasts, macrophages and lymphocytes. (Illustration: Per Stenqvist.)

exchange of O_2 and CO_2. The pulmonary vascular system, i.e. the capillary endothelium, also provides an important part of the gas exchange. The type II cells (pneumocytes) synthesise and secrete the alveolar surfactant that decreases the surface tension and prevents alveolar collapse. Following alveolar damage, the type II cells proliferate in order to regenerate both type I and II cells. In the alveoli there are also interstitial cells such as fibroblasts that secrete collagen and elastin, as well as alveolar and interstitial macrophages. Activated alveolar macrophages may contribute to toxicant-induced lung disorders. The lungs are enveloped by the pleura, a thin serous membrane.

Factors contributing to respiratory toxicity

Following inhalation of airborne agents, the nasal passages, conducting airways and lung alveoli will receive a high dose of the compound. Exercise may enhance the pulmonary dose because it increases ventilation and causes a switch from nasal to mouth breathing. The effects of inhaled gases, particles, drugs and chemicals depend on the sites of deposition, absorption and metabolism within the respiratory tract. Because the respiratory system has a large absorptive surface area and is highly vascularised, inhalation of drugs and chemicals may also lead to a rapid and extensive systemic exposure.

Because rats and mice are *obligate* nose breathers, nasal toxicity during inhalatory exposure is usually more pronounced in these species than in humans, who can also inhale through the mouth.

Deposition of gases and particles

Highly water-soluble gases such as sulfur dioxide (SO_2), ammonia (NH_3) and chlorine (Cl_2) are rapidly absorbed in the nasal passages or conducting airways and preferentially damage this site. The content of water-soluble irritants in inhaled air thus decreases in the conducting airways and little reaches the alveoli. Gases with low water solubility such as ozone (O_3) and nitrogen dioxide (NO_2) are poorly absorbed in the nasal passages, so that these gases can reach the alveoli and cause toxic responses at this site.

The sites of deposition of inhaled particles are determined by size and shape as well as the pattern of airflow in the airways. Large particles (>10 μm) are deposited in the nasal passages, whereas smaller particles (0.1–3 μm) are usually found in the conducting airways and alveoli. Inhaled ultrafine (nano)particles (<0.1 μm) will reach the alveoli and may elicit inflammatory responses in the lung with subsequent systemic effects. Ultrafine particles are found in diesel exhausts, but it is unclear whether the particle itself or environmental agents adsorbed on the particle are responsible for the harmful effects. The different sites of deposition of particles in the respiratory system can also be utilised for the site-specific pulmonary delivery of inhaled drug particles.

Soluble particles are dissolved in the mucus, absorbed by the epithelium and subsequently removed by the blood circulation. Insoluble particles trapped in the nasal and bronchial mucus are usually removed by mucociliary clearance mechanisms (Figure 9.1). The cilia in the nasal passages and conducting airways propel mucus with the trapped particles towards the glottis, where the mucus is swallowed and enters the gastrointestinal tract. In damaged airways such as metaplastic, nonciliated epithelia in tobacco smokers, the mucociliary clearance is less efficient. The cilia are not found beyond the terminal bronchioles and particles trapped in the alveoli may be removed by phagocytosis by alveolar macrophages. Depending on the nature of the particle, the macrophages may digest the material or not. Most alveolar macrophages are carried to the bronchioles by the alveolar fluid and removed by mucociliary clearance, while others are removed via the blood circulation or lymphatic drainage or are retained in alveolar walls for long periods. Activated macrophages can injure the pulmonary tissue via release of reactive oxygen species (ROS) and lysosomal enzymes.

Cardiac output and pulmonary accumulation of drugs and chemicals

The lungs receive the entire cardiac output of blood from the right ventricle and compounds that are injected subcutaneously or are applied topically pass through the lung before they reach other tissues. Pulmonary uptake may be involved in first-pass clearance and reduced bioavailability of drugs given by these routes of administration. There are efficient pulmonary uptake mechanisms for basic amines such as amiodarone, imipramine, selective serotonin-reuptake inhibitors (SSRIs) and local anaesthetics, and the lungs may function as a reservoir. The basic amines are accumulated in the lung by transporters in the pneumocytes or by specific binding sites in the pneumocytes. Transporters are proteins involved in the transport of specific substances across the membranes.

Metabolism of drugs and chemicals in the respiratory system

The expression of drug-metabolising enzymes in the human lung is generally low and there is usually only limited drug metabolism at this site. The pulmonary biotransformation may, however, lead to a first-pass clearance and reduced bioavailability of some drugs delivered via inhalation or by nasal administration. A wide range of drug-metabolising enzymes is present in the lining of the nasal and tracheobronchial mucosa, particularly in epithelial cells and glands. The cytochrome P450 (CYP) enzymes are often highly expressed in bronchiolar nonciliated cells (Clara cells) and in the nasal olfactory glands but also in type II pneumocytes. Furthermore, some CYP forms are expressed in pulmonary capillary endothelial cells. The high expression of CYP and other drug-metabolising enzymes in the nasal olfactory mucosa is most likely related to the clearance of odorants but may also protect the central nervous system against inhaled toxicants. The major pulmonary drug-metabolising forms of CYP are 2A13, 2B6, 2E1, 2F1, 3A5 and 4B1. Tobacco smoke is known to induce the carcinogen-metabolising CYP1A1 and 1B1 in human bronchial mucosa, alveoli and pulmonary vasculature via an aryl hydrocarbon (A*h*) receptor-mediated mechanism. Owing to the cellular heterogeneity in the respiratory tract, metabolism studies using whole lung or lung homogenates will normally not show the level of metabolism occurring in the metabolically active cell type(s). The metabolic activity in cell types with high expression of various CYPs will be diluted by the low activity in other cell types with low expression of CYPs.

The high expression of CYP and other drug-metabolising enzymes in specific cells and tissue structures in the respiratory system such as the nonciliated bronchiolar Clara cells and nasal olfactory glands makes these sites sensitive to chemicals that are bioactivated into electrophilic intermediates. The reactive metabolites may become covalently bound to nucleophilic sites, such as thiols present in proteins (protein adducts) or in DNA bases (DNA adducts). The covalent binding to tissue macromolecules is often regarded as an initial event leading to a cascade of biochemical changes and cell-specific damage. Notably, damage to the respiratory tract may also arise after exposure to blood-borne toxicants requiring metabolic activation.

Nose–brain transfer

The olfactory epithelium covers a minor area in the posterior part of the nasal passages and may provide a portal of entry for drugs and chemicals to the brain. The olfactory neurons have dendrites projecting into the nasal mucus and axons projecting into the olfactory bulb. The axons do not make synaptic connections until they reach the olfactory bulb. Inhaled metals such as manganese, but also intranasally administered drugs and chemicals, may be transferred via olfactory neurons into the olfactory bulb and in some cases further into the brain, thus by-passing the blood–brain barrier.

Toxicant-induced responses of the respiratory system

Although drug-induced adverse reactions are relatively rare, epidemiological studies have demonstrated significantly increased risks for acute or chronic responses in the respiratory system following exposure to tobacco smoke and other airborne pollutants.

Irritation

Acute symptoms associated with inhalation of irritant gases may include runny nose, nasal bleeding, sore throat, changed breathing pattern and cough. Severe irritation results in bronchoconstriction, mucus secretion, wheezing and shortness of breath. The effects of irritant gases in the respiratory tract may involve stimulation of the parasympathetic nervous system, leading to nasal secretion.

Acute respiratory distress syndrome

Acute respiratory distress syndrome (ARDS) may be related to toxicant-induced changes in the alveoli involving both the type I pneumocytes and capillary endothelial cells. Extensive damage to the capillary endothelium leads to increased permeability, resulting in accumulation of fluid, initially as interstitial oedema and subsequently as alveolar oedema. Inhalation of high levels of O_3, Cl_2 or smoke and overdosing of tricyclic antidepressants or salicylates may induce ARDS-like reactions. Commonly abused drugs such as opiates, cocaine, amfetamines and benzodiazepines may also lead to acute respiratory failure. *Primary pulmonary oedema* (noncardiogenic oedema) is a characteristic feature of ARDS, whereas cardiogenic pulmonary oedema is usually related to other disorders such as congestive heart failure. The leakage of fluid into the alveoli may overwhelm the lymphatic drainage and inactivate the surfactant, leading to impaired gas exchange and a life-threatening shortage of oxygen. The acute damage may resolve by epithelial regeneration and reabsorption of the oedematous fluid, or it may progress to pulmonary fibrosis.

Pulmonary fibrosis

Pulmonary fibrosis is characterised by the accumulation of connective tissue replacing the respiratory tissue. Chronic administration of some drugs and occupational exposure to irritant gases and particles can cause this disease. Activated macrophages and neutrophils may release cytokines and growth factors, leading to proliferation of fibroblasts. This is accompanied by an increased amount and changed composition of extracellular collagen in the lung. The lung becomes small and stiff and the gas exchange will be impaired. *Pneumoconiosis*, such as silicosis and asbestosis, is related to repeated inhalation of dusts, particles and fibres that are permanently deposited in the lung and induce pulmonary fibrosis. Inhalation of asbestos fibres may also induce lung cancer as well as *mesothelioma*, a malignant neoplasm arising from the pleura. Mesothelioma leads to a compressed lung surrounded by a thick layer of tumour. Heavy smokers exposed to asbestos have an increased risk of developing lung cancer, whereas smoking alone does not increase the risk of developing mesotheliomas.

Chronic obstructive pulmonary disease

Caused by chronic bronchitis or pulmonary emphysema, chronic obstructive pulmonary disease (COPD) is a serious condition of chronic irreversible airflow obstruction (decreased lumen diameter) and a diminished ability to expire air. It is characterised by a severe and finally life-threatening shortage of oxygen. Tobacco smoking is a major cause of COPD (>90% of reported cases). More than 4000 chemicals, including many pulmonary irritants and toxicants, are present in tobacco smoke, but the precise chemicals involved in the induction of COPD have not been identified. *Chronic bronchitis* is characterised by productive cough and excessive production of mucus as well as mucociliary dysfunction. There is an increased number (hyperplasia) and enlargement (hypertrophy) of mucus glands and goblet cells and the activity of the cilia is impaired. *Pulmonary emphysema* is characterised by gradual destruction of the alveolar walls, a decrease in elastic fibres (elastin), and a subsequent enlargement of the air spaces. In tobacco smokers there is an increased amount of neutrophils and macrophages in the respiratory system. The activated neutrophils and macrophages may release proteases such as elastase. In addition, free radicals in tobacco smoke are known to inactivate a pulmonary elastase inhibitor (α_1-antitrypsin). The resultant protease–antiprotease imbalance leads to enhanced breakdown of the alveolar wall.

Asthma

Asthma is an inflammatory disease characterised by airway hyperreactivity with spasmodic contraction of the bronchi in response to various stimuli, including allergens. The airflow obstruction is largely reversible. The bronchial inflammation is caused by type 1 hypersensitivity reactions involving interaction of an allergen with IgE antibodies (Chapter 10). Endogenous and exogenous sources of free radicals and reactive oxygen species in, for instance, tobacco smoke may also contribute to the inflammatory response. This condition can be triggered by airborne pollutants such as O_3 and SO_2 and in rare cases by NSAIDs (such as aspirin (acetylsalicylic acid)), beta-blockers and calcium antagonists. Toluene diisocyanate (TDI), used in the production of polyurethane plastics, may provoke occupational asthma. TDI reacts with proteins to form adducts that are recognised as an antigen by the immune system. Tobacco smoking is known to aggravate asthma and children of tobacco-smoking parents have increased risk of asthma.

Phospholipidosis

Phospholipidosis is a disturbance in phospholipid metabolism due to inhibition of the lysosomal phospholipase activity. This may lead to an abnormal deposition of lipids in alveolar cells and macrophages. Drug-induced phospholipidosis in the lung, brain and kidney can be detected in preclinical toxicity studies of new drug candidates.

Lung cancer

Lung cancer is a common malignant neoplasm that usually arises in the epithelium of the conducting airways. This neoplasm (*bronchial carcinoma*) is usually preceded by progressive changes including metaplasia or dysplasia in the bronchial/bronchiolar epithelium. There is a stepwise accumulation of oncogenic mutations in the epithelial cells and the p53 gene is mutated in about 60% of lung cancer patients. There are four major histological types of bronchial carcinomas. Small-cell carcinoma is a highly malignant neoplasm that metastasises to distant sites and is usually treated with chemotherapy. The non-small-cell carcinomas, including squamous cell carcinoma, adenocarcinoma (mostly in the peripheral lung of women) and large-cell carcinoma, are usually treated with surgery. The major cause of lung cancer is tobacco smoking and 80–90% of reported cases are found in smokers. More than 70 chemical carcinogens are present in tobacco smoke or tar and some of these compounds are known or probable human carcinogens according to the IARC assessment (see Chapter 6). Many of the tobacco-related carcinogens, such as tobacco-specific nitrosamines, can induce tumours in the respiratory tract of experimental animals. The risk of lung cancer is related to total lifetime exposure to tobacco smoke and even low-tar cigarettes are harmful. Furthermore, passive smoking causes a low but increased risk of developing this disease. The latency period for lung cancer is usually 20–40 years. Despite surgery, radiation and drug treatment, the 5-year survival rate is poor.

Methods for studying toxicity to the respiratory system

The effects of airborne drugs and chemicals on various pulmonary functions in humans and experimental animals can be examined by several methods *in vivo* and *in vitro*. Typical lung volume measurements such as forced expiratory volume in one second (FEV_1) and forced vital capacity (FVC) are described in detail in textbooks of human physiology. Standard toxicity testing in experimental animals usually includes morphological examination of the tissue by light microscopy. The determination of the wet weight of the lung may indicate the presence of pulmonary oedema. Biochemical and molecular changes can be examined in isolated perfused lung, in homogenates of the lung, in isolated cells or in lung lavage fluid. The lung lavage is performed by the administration of a sterile solution via the trachea. The resulting lavage fluid is often used for biochemical determination of enzyme activities such as that of lactate dehydrogenase (LDH) released into the alveolar spaces by damaged pneumocytes. To permit study of individual cell types in the heterogeneous respiratory tract, the tissue has to be dissociated and isolated cells separated.

Small rodents and dogs are frequently used for inhalation toxicity studies in order to determine local and systemic effects of drugs and chemicals. Notably, small rodents are obligate nose breathers, whereas humans and dogs can inhale air through both the mouth and the nose. The structure of the nasal passages is more complex in rodents than in humans and the olfactory mucosa occupies about a half of the total nasal area in rodents, whereas the olfactory mucosa is only about 3% of the total nasal area in humans.

Respiratory system toxicants

Drugs, chemicals and oxidative stress

The respiratory system is continuously exposed to oxygen and has a high oxygen tension. Oxygen is necessary for life; however, during the normal metabolism a stepwise reduction of O_2 with concomitant generation of reactive oxygen species (ROS) such as $O_2^{\cdot-}$, H_2O_2, and $OH^{\cdot}$ takes place (Chapter 3). Normally, the lung contains antioxidant defence systems including superoxide dismutase (SOD), glutathione (GSH), GSH peroxidase and catalase as well as alpha-tocopherol and ascorbic acid to decrease the oxidative damage. The antioxidant defence system can be overwhelmed, leading to formation of the highly reactive hydroxyl radical ($OH^{\cdot}$) and lipid peroxidation. In hyperoxia-induced lung injury the initial damage occurs in type I pneumocytes and in the capillary endothelium and the subsequent changes may lead to an ARDS-like reaction. The exposure of animals to 100% O_2, for instance, is fatal. Drugs and chemicals may elicit increased pulmonary oxidative stress by redox-cycling, i.e. shuttling of electrons to O_2, or by affecting the anti-oxidant defence system.

Bleomycins are a family of cytotoxic glycopeptide antibiotics that are used to treat testicular cancer and lymphomas. Chronic administration of this agent may induce endothelial damage in the pulmonary vasculature. This is followed by oedema and an influx of inflammatory cells. There is a release of cytokines, activating fibroblasts and leading to fibrosis. Bleomycins form a complex with Fe^{2+} that is oxidised to Fe^{3+} and reacts with oxygen. The activated bleomycins decompose to generate hydroxyl radicals. The specific sensitivity of the lung is considered to be due to absence of the main bleomycin-inactivating enzyme (bleomycin hydrolase) in the lung. In animals, the bleomycin-induced pulmonary toxicity and mortality are increased following exposure to high levels of oxygen. Oxygen supplementation is discouraged in human patients, although the effects of oxygen supplementation on bleomycin-induced human lung reactions have not been fully established.

Carmustine (BCNU) is an alkylating nitrosourea derivative used to treat lymphoma and other cancers. High doses are known to induce pulmonary toxicity and fibrosis that in some cases are fatal. The mechanism of pulmonary toxicity has not been clarified, but BCNU is known to inhibit lung GSH reductase, leading to a changed intracellular glutathione/glutathione disulfide (GSH/GSSG) ratio and less protection against oxidative stress.

Nitrofurantoin is a urinary antibacterial agent that is associated with fatal lung reactions. Acute reactions include laboured breathing and cough, whereas interstitial pneumonitis and fibrosis are common in chronic reactions. The pulmonary toxicity is related to transfer of an electron from NADPH and NADPH-cytochrome P450 reductase to the nitro group (Figure 9.2). The nitro radical interacts with O_2, resulting in redox-cycling of the compound with the concomitant generation of O_2^- and other ROS. Nitrofurantoin is not accumulated in the lung and the reason for the preferential damage in the lung has not been fully clarified. The high oxygen tension in the lung may play a role.

Accidental oral intake or incorrect use of the herbicide *paraquat* may cause fatal pulmonary toxicity. The initial damage occurs in type I and type II pneumocytes. Within a few days, secondary changes such as pulmonary haemorrhage and oedema will develop, followed by respiratory failure or delayed, fatal fibrosis. There is no effective antidote to acute poisoning, but a rapid haemoperfusion treatment may initially be used for these patients. The pulmonary toxicity is related to the active uptake of paraquat by the polyamine transporter in the pneumocytes.

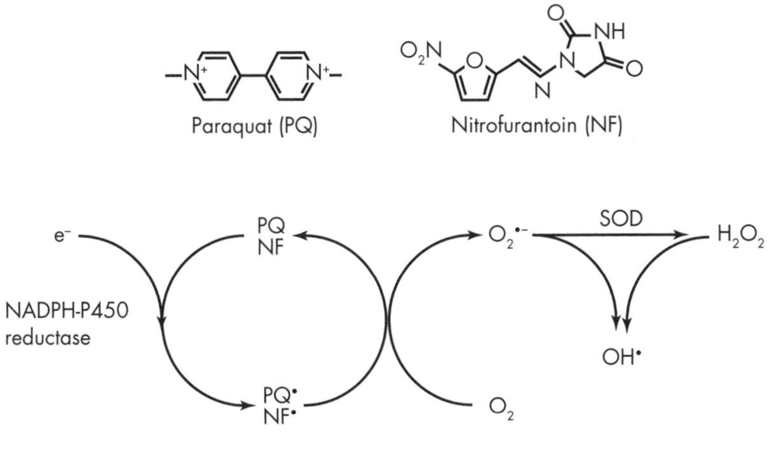

Figure 9.2 Redox-cycling of nitrofurantoin (NF) and paraquat (PQ). NF and PQ radicals are formed following transfer of an electron from NAPDH-P450 reductase. The NF and PQ radicals transfer the extra electron to oxygen, forming a superoxide anion radical with the concomitant regeneration of the parent compounds. The superoxide anion radical is spontaneously detoxified to hydrogen peroxide or metabolised by superoxide dismutase (SOD). A reaction between the superoxide anion radical and hydrogen peroxide may produce the highly reactive hydroxyl radical (Haber–Weiss reaction). The hydrogen peroxide may also be cleaved to a hydroxyl radical (Fenton reaction).

Paraquat is not metabolised or bound to the lung, but it forms a radical species following a transfer of an electron from NADPH and NADPH-cytochrome P450 reductase (Figure 9.2). The paraquat radical interacts with O_2, resulting in redox-cycling of the compound with the concomitant generation of $O_2\cdot^-$ and other ROS.

Toxicants requiring bioactivation in the respiratory tract

Ipomeanol

Ipomeanol is a cytotoxic natural product. This compound induces selective necrosis of the nonciliated Clara cells in the bronchiolar epithelium of several mammalian species. Ipomeanol is metabolically activated by rodent CYP4B1 to a reactive furan epoxide metabolite. Autoradiography shows that the reactive metabolite is covalently bound to the nasal and tracheobronchial epithelium. The reactive metabolite is detoxified by conjugation to GSH and toxic doses of ipomeanol decrease the level of GSH in the lung. Prior treatment with GSH-modulating agents, or with CYP inhibitors, influences both the level of the covalent binding and the pulmonary toxicity, suggesting that covalent binding of ipomeanol to lung proteins is a primary event in the mechanism of toxicity. Early clinical trials of the use of ipomeanol as a prodrug to treat lung cancer were not successful and revealed that human pulmonary CYP4B1 showed negligible activity towards ipomeanol. In contrast, this compound was hepatotoxic in humans and was activated by other CYPs in the liver.

Chlorinated and aromatic hydrocarbons

Repeated occupational or environmental inhalation of volatile hydrocarbons such as tetrachloroethylene and naphthalene may damage the respiratory tract. These compounds are bioactivated by various CYP forms that are present in the tracheobronchial epithelium. This leads to the formation of reactive, electrophilic intermediates and results in protein and/or DNA adducts.

Polyaromatic hydrocarbons

Polyaromatic hydrocarbons (PAHs) are well-known components of tobacco smoke and tar. These carcinogens are metabolically activated by CYP1A1 and CYP1B1 as well as the CYP3 family to epoxide intermediates, which are converted with the aid of epoxide hydrolase to the ultimate carcinogens, diol-epoxides that bind covalently to DNA (Figure 9.3). Human bronchial mucosa and peripheral lung cells can metabolise these carcinogens. DNA adducts can be detected both in the bronchial mucosa and in the peripheral part of the lung of smokers. A characteristic spectrum of p53 mutations in lung cancers from smokers has been attributed to the mutagenic action of PAHs. Experimental studies have demonstrated that the diol-epoxide metabolites of one model PAH (benzo[a]pyrene) preferentially form adducts to the most frequently mutated guanine nucleotides within p53 codons. Mutations in the p53 gene make cells more vulnerable to additional DNA damage.

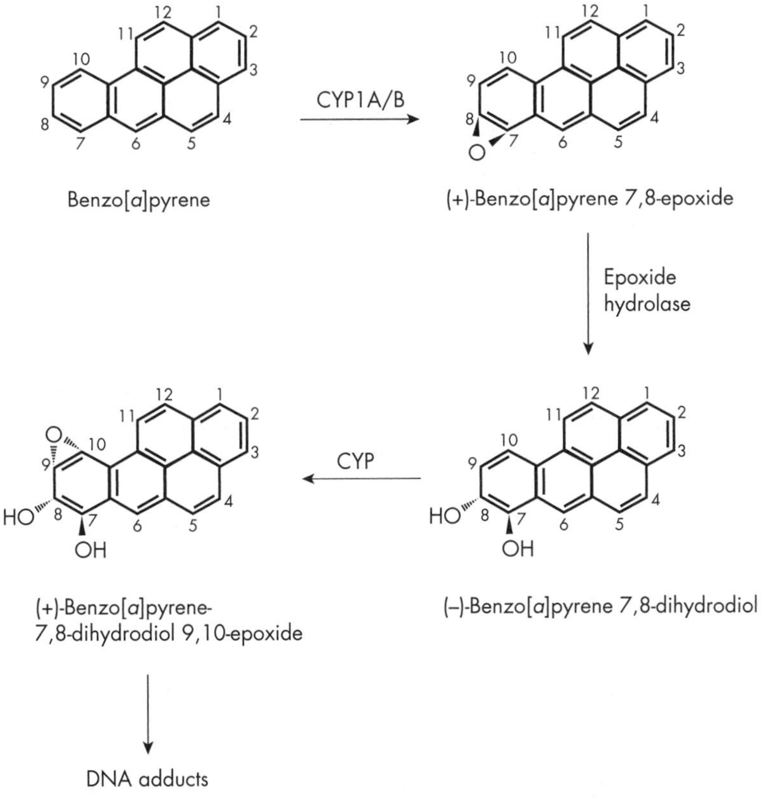

Figure 9.3 Metabolic activation the tobacco-related carcinogen benzo[a]pyrene (B(a)P). B(a)P is metabolised by CYP1A1/1B1 to epoxide intermediates, which are detoxified by epoxide hydrolase. The B(a)P 7,8-dihydrodiol 9,10-epoxide is resistant to metabolism by epoxide hydrolase and binds covalently to DNA.

N-Nitrosamines

More than 30 years ago, tumours in the nasal passages and lung were reported in rodents following systemic exposure to a number of N-nitrosamines. These potent carcinogens are now known to be extensively bioactivated by nasal olfactory and pulmonary CYP forms. The tobacco-specific N-nitrosamine N'-nitrosonornicotine (NNN) is α-hydroxylated by various forms of CYP2 enzymes (Figure 9.4). The α-hydroxylated nitrosamines are spontaneously degraded to reactive carbonium ions that bind covalently to DNA and proteins. Human trachea, bronchial mucosa and peripheral lung cells can metabolise these carcinogens and DNA adducts have been detected in human tissues. In rodents, a preferential localisation of adducts has been detected in the nasal and tracheobronchial mucosa (Figure 9.5).

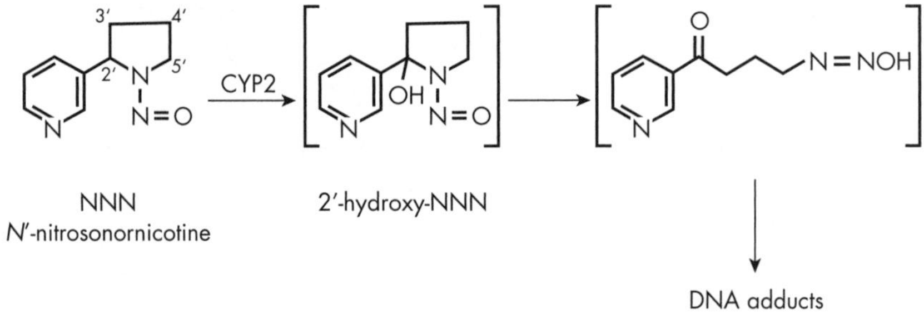

NNN
N'-nitrosonornicotine

2'-hydroxy-NNN

DNA adducts

Figure 9.4 Metabolic activation of the tobacco-related carcinogen N'-nitrosonornicotine (NNN). NNN is α-hydroxylated by CYP2 enzymes. The α-hydroxylated NNN is spontaneously degraded to a reactive carbonium ion that binds to DNA and proteins.

Heterocyclic amines

A number of heterocyclic amines are often called food mutagens because they are formed from amino acids in the cooking of meat and fish. These carcinogens are also formed in the burning of tobacco. In humans these compounds are preferentially N-hydroxylated by CYP1A2. The N-hydroxylated metabolites may react directly with DNA, but they are often conjugated by N-acetyltransferases or sulfotransferases to form highly reactive N-acetoxy or N-sulfonyloxy esters that readily react with DNA. Human lung can metabolise tobacco-related heterocyclic amines.

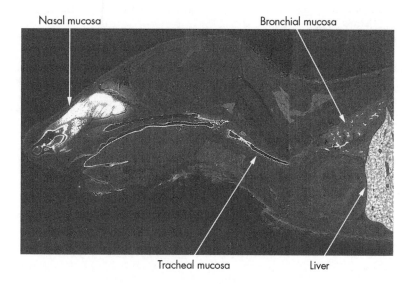

Nasal mucosa   Bronchial mucosa

Tracheal mucosa   Liver

Figure 9.5 Autoradiography of the tobacco-related carcinogen N'-nitrosonornico-tine (NNN). A rat was injected intravenously with ^{14}C-labelled NNN and killed one day later. The autoradiogram shows selective retention of radioactive adducts in the mucosa of the nasal region, trachea, bronchi and bronchioli and in the liver. White areas correspond to high levels of radioactivity.

Cationic amphiphilic drugs

Cationic amphiphilic drugs (CADs) share common physicochemical properties and often induce phospholipidosis in rodents. However, only a few drugs, such as the antiarrhythmic amiodarone and the anti-depressant fluoxetine, are known to induce this condition in humans. Amiodarone treatment is associated with cough and difficult breathing. The CADs accumulate in lysosomes and inhibit the phospholipase activity, resulting in the accumulation of phospholipids within the organelle. The CAD-induced pulmonary phospholipidosis usually disappears upon termination of exposure.

Further reading

Ben-Noun L (2000). Drug-induced respiratory disorders: incidence, prevention and management. *Drug Safety* 23: 143–164.

Foster JR (1989). The respiratory system. In: Turton J, Hooson J, eds. *Target Organ Pathology. A Basic Text*. London: Taylor & Francis, 335–369.

Foth H (1995). Role of lung in accumulation and metabolism of xenobiotic compounds – implications for chemically induced toxicity. *Crit Rev Toxicol* 25: 165–205.

Gram TE (1998). Chemically reactive intermediates and pulmonary xenobiotic toxicity. *Pharmacol Rev* 49: 297–341.

Hukkanen J, Pelkonen O, Hakkola J, Raunio H (2002). Expression and regulation of xenobiotic-metabolizing cytochrome P450 (CYP) enzymes in human lung. *Crit Rev Toxicol* 32: 391–411.

Limper AH (2004). Chemotherapy-induced lung disease. *Clin Chest Med* 25: 53–64.

10

Immunotoxicity

Camilla Svensson

The primary mission of the immune system is to rid the host of invading infectious agents. This is effectively managed through finely tuned actions of a network of different cell types, cellular products and organs, and is highly dependent on the ability to recognise and distinguish between self and non-self. Drugs can interfere with this system with severe consequences for the patient; a loss of response to foreign antigens may lead to increased susceptibility to infections and certain types of cancer, while a loss of tolerance to self-antigens can manifest in autoimmune diseases.

Drugs (drug adducts) can also be direct targets of an immune response and give rise to different kinds of hypersensitivity reactions. Such reactions are very difficult to predict in animal experiments, and there are several examples of drugs that had to be withdrawn owing to the appearance of severe drug reactions in some patients.

The immune system

The core components of the immune system are the B- and T-lymphocytes (B- and T-cells), which collectively express an almost unlimited repertoire of different antigen receptors (B-cell and T-cell receptors) that enables them to detect and mount a specific immune response to virtually any foreign agent (antigen) that enters the body. This response also leads to the formation of memory cells that, upon subsequent encounters with the antigen, react swiftly to remove the antigen. T-cells also have the ability to distinguish between self and non-self, which protects the host's own tissues from being attacked by its own cells.

The *specific immunity* provided by lymphocytes is complemented by cells and soluble factors of the *innate immunity*, which includes macrophages, neutrophils, eosinophils, dendritic cells, NK (natural killer) cells and complement factors. These components can eliminate large numbers of antigens through phagocytosis, lysis or induction of apoptosis, which is important both as a first line of defence and in the late phase of a specific immune response when these components are recruited by T-cells and B-cell-secreted antibodies to the site of the antigen. In addition, macrophages and dendritic cells are critical for the activation of the specific immune response by being able to present antigens to T-cells. For a thorough background on the immune system, the reader is referred to a textbook on immunology.

Immunosuppression

Immunosuppression is the summary term for a decreased ability of the immune system to mount an immune response to a foreign antigen. It manifests as an increased susceptibility to infections and increased risk for certain types of neoplasms such as skin and lip cancers and non-Hodgkin's lymphoma. Many immunosuppressive drugs are used therapeutically in modulating immune responses, e.g. to prevent transplant rejection or to treat autoimmune diseases.

Mechanisms behind immunosuppression

Because of the complexity of the immune system, immunosuppression can be generated by many different mechanisms acting directly or indirectly on one or several components of the immune system. Three particularly sensitive processes are exemplified below.

Cell proliferation

All cells of the immune system derive from self-renewing haematopoietic stem cells in the bone marrow. These and other proliferating cells in the immune system are well-known targets for anti-proliferative drugs (e.g. cyclophosphamide, methotrexate and azathioprine), which may lead to different types of anaemia. The same drugs can also suppress immune response by inhibiting clonal expansion of activated T- and B-cells.

The development of T- and B-cells

The development of T- and B-cells occurs in the specialised micro-environments of thymus and bone marrow, respectively, where lymphoid progenitor cells are stimulated to go through a complex series of rearrangements of antigen receptor genes, selections and proliferation steps. Potentially autoreactive cells are eliminated in this process and developing T-cells are also positively selected for their ability to recognise the host's major histocompatibility complex (MHC) molecules. An effect on lymphocyte development can lead to decreased numbers of functionally mature T- and B-cells, and thus inhibition of cellular and humoral immune responses. Examples of drugs that affect lymphoid development are glucocorticoids and ciclosporin (cyclosporin A).

T-cell activation

The most critical event in the specific immune response is the interaction of the T-helper cell with the antigen-presenting cell, which leads to the activation, proliferation and differentiation of the T-cell. In addition to the specific T-cell receptor–antigen–MHC interaction, a number of co-stimulatory receptors, adhesion molecules, secreted cytokines and intracellular signal transduction molecules participate in the stimulation of the T-cell.

Any drug that interferes with either expression or function of a molecule in this chain of events could potentially disturb the initiation of an immune response. A number of immunosuppressive drugs act on this activation step: ciclosporin, sirolimus (rapamycin) and muromonab-CD3 (OKT3), a monoclonal antibody that blocks the T-cell receptor. All these drugs are used in preventing transplant rejections.

Immune enhancement

The reverse effect of immunosuppression is the situation in which a drug enhances an immune response. There are a few examples of drugs that are used therapeutically to restore immune competence (e.g. levamisole). Common adverse effects of immune enhancement treatments are fevers, chills and hypotension. In rare cases, potentiation of asthma and eczema has been observed.

Hypersensitivity

The main function of the immune system is to recognise and eliminate foreign material. With this in mind, it is not so strange that drugs themselves can be targets of an immune response.

Hypersensitivity occurs when an unwanted and exaggerated immune response is directed towards an otherwise harmless antigen. The response is sometimes so strong that it causes severe tissue damage or even death. Of particular concern is that drug-induced hypersensitivity reactions are highly idiosyncratic and difficult to predict. Hyper-sensitivity reactions can be separated into four different types, which differ in the immunological mechanisms leading up to tissue damage. Reactions of types I–III are mediated by antibodies, which recruit complement or inflammatory cells to the site. Type IV reactions are mediated by antigen-specific T-cells (T_H1 cells), which engage macrophages, cytotoxic T-cells or eosinophils in the effector response (Figure 10.1).

Development of hypersensitivity to a drug generally requires at least two separate exposures. The initial exposure, called the sensitisation phase, induces a primary response when the antigen is presented to

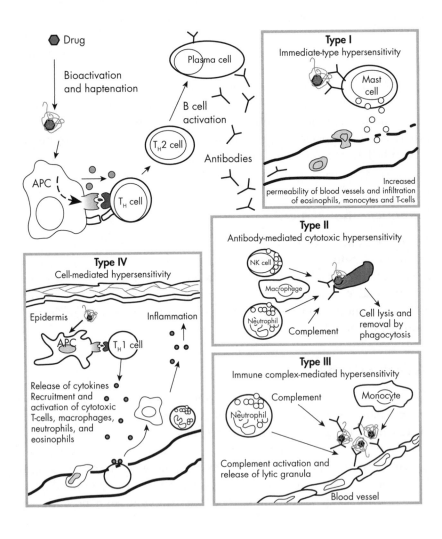

antigen-specific T-helper cell by antigen-presenting cells. The activated T-helper cell can then differentiate to either a T_H1 cell, which promotes T-cell-mediated responses, or to a T_H2 cell, which stimulates antigen-specific B-cells to produce antibodies. The primary response usually does not generate any symptoms, in part because the antigen may already have been degraded/metabolised by the time effector cells start to appear (7–10 days later). However, memory cells and antigen-specific antibodies that have formed will be ready to engage in a rapid and vigorous response (referred to as the elicitation phase) next time the antigen appears. Even though only one exposure is necessary to elicit sensitisation, hypersensitivity most often occurs after long-term exposure.

Figure 10.1 Hypersensitivity reactions.

Type I: immediate-type hypersensitivity (allergy). Antigen cross-linking of mast cell-bound IgE antibodies releases preformed mediators such as histamine and induces the production of other inflammatory mediators (e.g. cytokines and leukotrienes). Within minutes, this leads to dilatation and increased permeability of local blood vessels and constriction of smooth muscles. Symptoms range from rashes, rhinitis and asthma to systemic life-threatening anaphylaxis. Examples: penicillins, streptokinase, trimethoprim.

Type II: antibody-mediated cytotoxic hypersensitivity. An IgG- (or IgM-) mediated response is directed towards a cell-bound antigen. The cell is damaged by complement activation or the cytotoxic action of macrophages, natural killer (NK) cells and neutrophils that are recruited by the antibodies to the cell surface. Erythrocytes and thrombocytes are examples of target cells in this rare type of hypersensitivity reaction, which results in haemolytic anaemia and thrombocytopenia, respectively. Examples: penicillin and methyldopa.

Type III: immune complex-mediated hypersensitivity. Type III reactions occur when (immune) complexes between soluble antigens and IgG antibodies are not cleared away efficiently by phagocytosing cells. The immune complexes can then diffuse away and deposit at various sites in the body (e.g. capillaries, the glomeruli of the kidneys and synovial membranes), which are damaged by complement activation and cytotoxic cells. Manifestations of type III reactions include systemic lupus erythematosus and serum sickness. Example: penicillins.

Type IV: cell-mediated hypersensitivity (delayed-type hypersensitivity). Type IV reactions are orchestrated by sensitised T_H1 cells, which upon contact with antigen-presenting cells (APC) secrete cytokines that support activation of cytotoxic T-cells and/or recruit inflammatory cells (e.g. macrophages, eosinophils) to the site of exposure. Type IV reactions are often the result of topical (contact hypersensitivity) or intradermal (tuberculin-type hypersensitivity) administration of drugs and result in symptoms such as eczema and erythema 24–72 hours after antigen exposure. Examples: penicillins, sulfonamide, lidocaine.

Mechanisms behind drug-induced hypersensitivity

Development of hypersensitivity is a multistep process that can be influenced by several factors. This section will follow the steps that lead up to hypersensitivity and some important parameters that influence each of these processes.

Formation of hapten–carrier complexes

The immune system typically reacts to larger molecules (>1000 Da), so that most drugs are not immunogenic per se. In order to generate an immune response, a drug must therefore conjugate with larger molecules such as serum or cellular proteins, forming what is called a hapten–carrier complex. This complex can then be processed by antigen-presenting cells and can be recognised as foreign by specific T-helper cells.

Haptenation can only occur if the compound is chemically reactive. Some drugs such as penicillins and cephalosporins, although not very reactive in general, can conjugate directly with certain proteins (see Figure 10.2a). However in most cases of drug hypersensitivity, the reactivity is generated through bioactivation of the drug.

As discussed in Chapter 2, bioactivation is most active in the liver but, thanks to the presence of highly efficient detoxifying systems and poor immunological competence, the liver is rarely targeted by adverse immune reactions (autoimmune hepatitis caused by the diuretic tienilic acid and halothane being some of the exceptions). Instead, it has been shown that extrahepatic metabolism can play an important role in haptenation. For instance, the skin, bone marrow and respiratory tract all have appreciable drug-metabolising capacity combined with immunological competence. A drug that is absorbed by, or distributed to, these organs can be taken up by macrophages, neutrophils or Langerhans cells (a dendritic cell type present in skin) and be metabolised by enzymes of the cytochrome P450 family and the myeloperoxidase/NADPH oxidase system. If a reactive product forms, it can directly conjugate with cellular proteins in these cells and be presented to T-cells.

Factors that influence the formation of hapten–carrier complexes include chemical properties of the drug, how it is administered and genetic polymorphism in drug-metabolising genes (e.g. slow/fast acetylators). Moreover, the counterbalancing action of detoxification systems such as glutathione conjugation can also be limited. Stress, drug interactions, and ongoing infections may all increase the risk of sensitisation due to reduced elimination of reactive metabolites.

The pharmacological interaction (p-i) concept

The p-i concept is a hypothesis that explains the fact that certain drugs can activate antigen-specific T-cells independently of bioactivation and antigen processing (see Figure 10.2b). These compounds can bind directly to peptides in the MHC cleft. The binding is of low affinity but sufficient to activate a T cell with a matching receptor. Examples of drugs that have been shown to cause hypersensitivity reactions through p-i are lidocaine, celecoxib, ciproxin and sulfamethoxazole.

Induction of danger signals

Even when a drug has been haptenated (or has bound directly to MHC molecules) and is being presented to a T-cell, it is not certain that an immune response will be initiated. This is because the immune system has developed tolerance mechanisms to avoid unwanted immune responses. Naive T-cells need to receive at least two different signals to become activated. One of these signals must always be the antigen-specific interaction between the T-cell receptor (TCR) and the peptide–MHC complex on the antigen-presenting cell (APC). The second signal is also delivered by the APC through co-stimulatory receptors such as B7 (which interacts with CD28 on the T-cell), but this second signal is only induced as a response to danger signals (e.g. inflammatory cytokines), the production of which is triggered by the presence of pathogens, cellular stress or physical damage (see Figure 10.2c). Thus a drug that causes general toxicity or is irritating is more likely to activate an immune response through this second signal. In the absence of a danger signal, the T-cell will most likely ignore the antigen and become tolerant to it.

Factors regulating T_H1 vs T_H2 responses

Once a T-cell has been activated, it can follow different effector pathways. It can either develop into a T_H1 cell, which promotes cell-mediated (Type IV) reactions through stimulation of macrophage and cytotoxic T-cell activity, or it can differentiate into a T_H2 cell, which has the ability to activate B-cells and thus promote antibody-mediated reactions (Types I–III).

The choice between a T_H1 and a T_H2 response is regulated by several factors, but the most critical is the types of cytokines that are released locally by other immune cells. For instance, activated dendritic

Benzylpenicillin

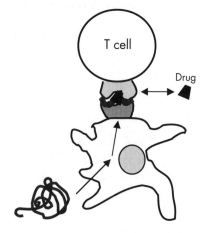

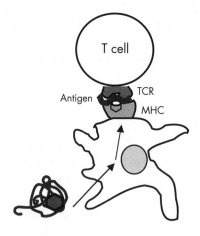

(a)   Benzylpenicilloyl-protein conjugate

Classical presentation of drug conjugates: Pharmacological interaction:
Metabolism/processing dependent   Metabolism/processing independent

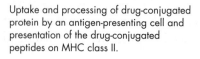

Uptake and processing of drug-conjugated   Low affinity interaction of drug
protein by an antigen-presenting cell and   with peptide on MHC surface.
presentation of the drug-conjugated
(b)   peptides on MHC class II.

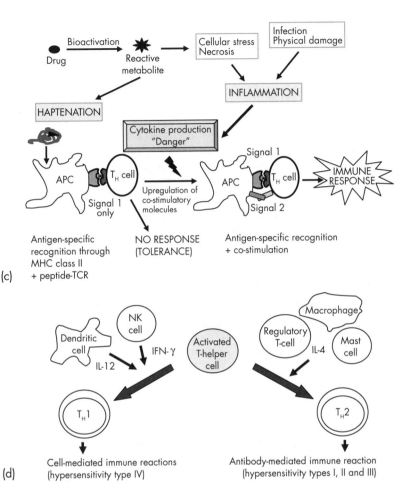

Figure 10.2 (a) *Formation of hapten–carrier complexes.* The β-lactam ring in benzylpenicillin is unstable and readily opens up to expose a carbonyl group, which then reacts with amines in lysine residues in nearby proteins to form a so-called hapten–carrier complex.

(b) *The pharmacological interaction p-i concept.* An immune response to a drug usually occurs due to metabolism-dependent hapten–carrier formation, followed by uptake, processing and presentation by antigen-presenting cells to T-cells (right). Some drugs can also bind, usually with low affinity, directly to the peptide–MHC complex on the surface of the antigen-presenting cell and thereby activate specific T-cells (left).

(c) *Induction of danger signals.* Inflammation due to cellular stress, physical damage or presence of pathogens can lead to induction of co-stimulatory molecules on the antigen-presenting cells, which increases the risk of activation of an immune response directed towards the drug-conjugate.

(d) *Factors regulating T_H1 vs T_H2 responses.* The choice between a T_H1 versus a T_H2 fate is influenced by cytokines excreted locally by other immune cells.

cells and NK cells produce IL-12 and IFN-γ, which promotes T_H1 differentiation, while IL-4 produced by antigen-presenting cells, mast cells and regulatory T-cells is a strong inducer of T_H2 differentiation (see Figure 10.2d).

Cytokines expressed by T_H1 and T_H2 cells also influence downstream effector mechanisms. T_H1 secretes IL-2, which supports activation of cytotoxic T-cells and IFN-γ and TNF-β which stimulates macrophages. T_H2 cells produce cytokines such as IL-4, IL-5, IL-6 and IL-10, which are important for B-cell activation and the subsequent production of different isotypes of antibodies.

The type of antigen, the dose, the route of exposure and genetic factors of the host will influence what cytokine pattern comes to dominate. A low dose and inhalation of a drug usually promote IgE responses, while a high dose and oral administration lead to IgG responses. Type IV reactions are often the result of topical exposure. Genetic factors that influence T_H1 vs T_H2 responses are an increased tendency to mount IgE responses (so-called atopy), which is connected with polymorphism in regulatory regions upstream of a cluster of genes encoding the cytokines involved in class-switching, and genes coding for the high-affinity IgE receptor.

Autoimmunity

In hypersensitivity reactions the target of the immune response is the drug itself or a drug-conjugated protein. However, in drug-induced autoimmunity the immune response is diverted towards self-proteins, which may cause severe tissue damage and autoimmune-like diseases. Both autoreactive antibodies and T-cells can be involved in autoimmunity and the mechanisms of tissue injury can be classified by the same system (reaction types I–IV) as hypersensitivity reactions. The most frequent types of reactions are the Type II–IV reactions; i.e. the injury may develop from an antibody-mediated cytotoxicity directed towards the cells that express the autoantigen (Type II), as a consequence of a deposition of immune-complexes at different locations in the body (Type III) or due to the effects of a local inflammation response induced by autoreactive T cells (Type IV).

Autoimmune diseases can be organ-specific or systemic. In organ-specific diseases the self-antigen is restricted to a specific cell type or organ. Examples of this type are methyldopa-induced autoimmune haemolytic anaemia, where the response is directed against the Rh-molecule on erythrocytes, and halothane-induced autoimmune hepatitis (a response against CYP450 enzymes). In systemic autoimmune diseases, autoantibodies form against many different antigens, often ubiquitous intracellular molecules such as histone proteins or nucleic acids. Such antigens can be released from dying cells and, if they are not cleared away efficiently, can cause immune-complex-mediated reactions and widespread tissue damage. The classical example of systemic autoimmunity is systemic lupus erythematosus (SLE), which can be induced by drugs such as hydralazine, procainamide and isoniazid.

Drug-induced autoimmunity is a very rare side-effect that is often associated with long-term use and high doses of certain drugs. In most cases the symptoms regress upon removal of the drug.

Mechanisms behind autoimmunity

How can a drug generate and direct an immune response against self? First of all it must overcome the immunological tolerance mechanisms that prevents reactions to self-molecules. This can occur through the following processes.

1. *Interference with central tolerance.* Most autoreactive T- and B-cells are eliminated by negative selection before leaving the thymus and bone marrow. Drugs that disturb this elimination process may increase the risk of autoimmune reactions.

2. *Alteration of peripheral tolerance.* Some autoreactive cells may escape negative selection but, in a normal situation, their activation is prevented in the periphery since there is not enough co-stimulation from danger signals. Autoreactivity is also controlled in the periphery by regulatory T-cells, which secrete inhibitory cytokines such as IL-10 and TGF-β (transforming growth Factor β). A breach in peripheral tolerance can occur if a drug induces inflammatory signals, e.g. by causing tissue damage or cellular stress.

3. *Alteration of antigen processing and presentation of 'new' peptides.* Certain drugs, or reactive metabolites, may also alter proteins that are being processed by an APC so that a new antigen to which there is no tolerance is displayed. Reactive compounds can, for example, oxidise side chains of self-proteins so that they become immunogenic.

4 *Genetic predisposition.* Several genetic factors contribute to the development of drug-induced autoimmunity; for example, expression of certain MHC alleles and polymorphisms in genes

involved in the clearance of dead cells (complement factors, DNase I) and the phagocytosis of immune complexes.

5.   *Cross-reactivity*. A drug may elicit an immune response, which generates drug-specific T-cells and antibodies that cross-react with self-proteins.

Examples of immunomodulating drugs

Penicillins and cephalosporins

Antibiotics of the β-lactam group (penicillins and cephalosporins) provide the best-characterised examples of drug-induced hypersensitivity. Type I reactions are most common, but this group can give rise to all four classes of hypersensitivity reactions. Symptoms range from rashes, to haemolytic anaemia and potentially fatal anaphylaxis.

The main factor behind the immunogenicity of this group is the β-lactam ring itself, which can react with free lysine groups of proteins. Native β-lactams can form hapten conjugates directly, but metabolites such as penicillic acid are even more reactive and are the major cause behind sensitisation to these compounds. IgE antibodies formed against one type of penicillin can sometimes cross-react with other penicillins and even cephalosporins. Such effects are difficult to predict but can be identified by simple blood tests.

Procainamide

Procainamide has been used in the treatment of cardiac arrhythmias and myotonia since the 1950s. It has also long been known that the drug increases the risk of SLE-like autoimmunity. The symptoms of procainamide-induced SLE are usually milder in comparison to idiopathic SLE, e.g. skin changes and kidney damage (glomerular nephritis) are less common. Instead, pulmonary symptoms predominate, which could be due to local bioactivation by phagocytic cells. Anti-nuclear antibodies are detected in a majority of patients.

Long-term treatment (>3 months) at high doses is a risk factor, but several genetic factors have also been shown to increase the risk of this adverse effect.

• *Acetylator type*. Procainamide is mainly eliminated via acetylation. Consequently, slow acetylators are more prone to develop both hypersensitivity and autoimmune reactions to procainamide.

Inefficient hepatic acetylation is believed to increase bioactivation by neutrophils and macrophages.

- *MHC type*. Another genetic factor that increases the risk of SLE is expression of MHC class II molecules such as HLA DR4.
- *Complement system*. Defects in the complement system reduce the ability to clear away immune complexes. In addition, high concentrations of procainamide have been shown to directly inhibit activation of the complement system.

Glucocorticoids

Glucocorticoids are endogenously produced steroid hormones that have many important physiological functions. Their production is induced during immune responses, which they influence by exerting potent anti-inflammatory effects. Accordingly, synthetically produced glucocorticoids have become widely used as therapeutic agents in the treatment of different inflammatory and immune disorders.

Examples of effects induced by glucocorticoid treatment, for good and ill, are suppression of cell-mediated and humoral immune responses, inhibition of leucocyte migration to sites of inflammation, and deletion of developing thymocytes and activated T-cells. Most of the anti-inflammatory effects are mediated via inhibition of the transcription factor NFκB, which is an important activator of many immune response genes. For instance, NFκB is required for the production of many cytokines and their receptors (e.g. IL-2, IL-1β, TNF-α), chemotactic factors (e.g. IL-8) and adhesion molecules (e.g. vascular cellular adhesion molecule-1) upon cell stimuli.

Rapamycin

Sirolimus (rapamycin) is a potent immunosuppressive and antitumour agent and is one of the most commonly used agents in the prophylaxis against renal graft rejections. Sirolimus inhibits growth factor (e.g. IL-2) induced proliferation of T- and B-cells. It specifically binds to the immunophilin FK-binding protein and the formed complex blocks the activity of mTOR (mammalian target of rapamycin), a cell-cycle specific kinase. mTOR acts downstream of the IL-2 receptor in T-cells to promote progression from late G_1 to the S phase.

Immunotoxicity testing of drugs

The immunotoxic potential of a drug is evaluated using a combination of different tests that assess different components of the immune system. Standard toxicity testing protocols include a number of parameters, which can be evaluated in an initial screen for immunotoxic effect. If any abnormalities of immunocompetence are observed in these studies, more advanced methods can be applied to provide more detailed information on potential target cells and mechanisms of action. This tiered evaluation procedure also weights in other factors such as the severity of the effects, dose dependency, reversibility, and whether the compound or its metabolites are retained at high concentration in immune cells.

Assessment of basic immunological parameters using standard toxicity tests

Haematology and clinical chemistry   A complete count of blood cells is made and the different types of leucocytes are enumerated to detect specific haematological changes and possible target cells. Serum immunoglobulins are also analysed for drastic changes, which may provide evidence for immunosuppression or immune enhancement.

Examination or organ weights and histopathology   Thymus, spleen, bone marrow and lymphoid nodes that drain the site of drug administration are weighed and examined for gross signs of atrophy or hyperproliferation, which may indicate immunosuppression or immune enhancement, respectively. Histochemical evaluation of tissue morphology is also standard methodology.

Additional tests performed if signs of immunotoxicity are observed in standard testing

Immunophenotyping   A more detailed analysis of different subsets of immune cells, their activation state and localisation can be obtained using specific antibodies in combination with flow cytometry and immunohistochemistry.

Assessment of innate immunity   The activity of macrophages and NK cells isolated from drug-exposed animals can easily be assessed *in vitro*. Each cell type can be incubated with cells that have been preloaded with ^{51}Cr. The phagocytic activity of macrophages is measured by analysing the amount of radioactivity taken up by these cells when phagocytosing ^{51}Cr-labelled sheep red blood cells. Natural killer (NK) cell activity is measured by counting the radioactivity released to the culture medium after NK-cell-mediated lysis of ^{51}Cr-YAC-1 cells (a tumour cell line).

Assessment of humoral immunity   T-cell-dependent antibody production can be evaluated by the antibody-forming cell assay. Drug-exposed and control mice are injected intraperitoneally or intravenously with sheep red blood cells, which initiates the immune response. Blood is collected from the mice 5–6 days after immunisation and serially diluted into microtitre wells that have been precoated with the antigen (sheep red blood cells), to which formed antibodies will bind. The amount of antibodies formed can then be analysed using ELISA (enzyme-linked immunosorbent assay).

Assessment of cell-mediated immunity   Cell-mediated immunity can be evaluated *in vivo* by performing a delayed hypersensitivity response (DHR) test. The ability of T-cells to respond to an antigen by proliferation can also be measured *in vitro* by treating the cells with different agents such as anti-CD3 antibodies + interleukin-2 or phytohaemagglutinin. Another alternative is a mixed lymphocyte response test, in which T-cells isolated from drug-exposed animals are allowed to react with cells expressing different MHC molecules.

Host resistance studies   Different types of pathogens (e.g. *Listeria monocytogenes*, *Streptococcus pneumoninae* and influenza virus) or tumour cells can be used to challenge various immune defence mechanisms of drug-treated animals. These tests are mainly used as a final tier to confirm that a suspected immunosuppressant affects a specific target cell.

Assessment of hypersensitivity and autoimmunity

A drug's potential to induce hypersensitivity or autoimmunity is especially challenging to evaluate owing to the many different factors that influence these responses. Many of the methods available for testing the hypersensitivity potential of drug reactions focus on locally applied compounds.

Type I reactions are often studied in guinea pigs using the passive cutaneous anaphylaxis assay, in which the animal is intradermally injected with a test compound in order to provoke sensitisation. The animal is later challenged by an intravenous injection of the drug plus a tracer dye. If the compound causes a Type I reaction, there is a change in local vascular permeability, which results in increased extravascular staining by the dye.

Type IV reactions can be evaluated by many different methods such as the guinea pig maximisation test and the local lymph node assay. The guinea pig maximisation test measures the erythema and oedema formed after topical exposure of the drug to presensitised animals. The local lymph node assay is done on mice and analyses the primary T-cell response (i.e. antigen-induced proliferation) in draining lymph nodes following topical application of the test substance to the mouse ear.

There are presently no standard preclinical methods available to predict Type II and Type III hypersensitivity and autoreactivity, but such tests are under development. Autoimmune reactions are especially difficult to evaluate because they often appear after long-term exposure and can be difficult to differentiate from idiopathic reactions.

Further reading

Cavani A, Ottaviani C, Nasorri F, Sebastiani S, Girolomoni G (2003). Immuno-regulation of hapten and drug induced immune reactions. *Curr Opin Allergy Clin Immunol* 3(4): 243–247.

Goldsby RA, Kindt TJ, Osborne BA (2002). *Kuby Immunology*. New York: WH Freeman.

Naisbitt DJ, Gordon SF, Pirmohamed M, Park BK (2000). Immunological principles of adverse drug reactions: the initiation and propagation of immune responses elicited by drug treatment. *Drug Safety* 23(6): 483–507.

Putman E, van der Laan JW, van Loveren H (2003). Assessing immunotoxicity: guidelines. *Fundam Clin Pharmacol* 17(5): 615–626.

Vial T, Choquet-Kastylevsky G, Descotes J (2002). Adverse effects of immuno-therapeutics involving the immune system. *Toxicology* 174(1): 3–11.

11

Clinical toxicology

Hans Persson

This chapter deals with general aspects of clinical toxicology, with special emphasis on antidote treatment. Poisoning may be caused by chemicals, drugs and natural toxins. The focus is on poisoning by drugs, but chemicals and natural toxins will be addressed briefly when relevant with regard to toxic mechanisms, antidote treatment and elimination procedures.

What is clinical toxicology?

Clinical toxicology deals with diagnosis and treatment of poisoning, but also includes the study of mechanisms, toxicokinetics, toxicodynamics and the development of new treatment strategies. Clinical toxicology is rarely recognised as a medical specialty of its own. Depending on the type and severity, poisonings are mostly managed within internal medicine, paediatric, occupational medicine, ophthalmology, otorhino-laryngology and intensive care wards.

A number of clinics around the world have specialised in the management of poisoning, however. This has had a great impact on the development of evidence-based treatment principles in clinical toxicology. The establishment of special poisons information centres has also been of fundamental importance for the implementation of clinical toxicology. In relation to the significance of antidote treatment of poisonings, clinical toxicology also has an important link to pharmacy.

Poisons information centres

Because of the wide range of possible toxic exposures in modern society, special poisons information centres have been established. Sometimes these units are linked to treatment wards or laboratory facilities. Poisons information centres were gradually created in North America from the mid-1950s, in Europe from around 1960, and over the last few decades in the developing world also.

The main task of these units is to provide assessment and information on toxic risks, symptoms and adequate treatment in cases of acute poisoning. Surveillance of toxic events and preventive work is another important role of these centres.

General aspects

Acute and chronic poisoning

Poisonings can be acute or chronic, and a mixture of both forms can also occur. In acute poisoning, the body is exposed to the poison on one occasion and in a high dose. Toxic symptoms are often prominent and of rapid onset. However, symptoms may sometimes be delayed for many hours or even days; examples are poisonings by lithium, paracetamol (acetaminophen) and mushrooms containing amatoxins (*Amanita* species) or orellanine (*Cortinarius* species).

In contrast, chronic poisoning will follow repeated exposure to small amounts of the poison each time. Symptoms are seldom evident after each separate exposure; instead, the patient will become ill after weeks, months or even years. This may be due to a gradual accumulation due to specific binding, or to reduced capacity to eliminate or metabolise the 'poison'. Well-known examples are chronic intoxications by digitalis, lithium and heavy metals.

Toxic mechanisms

For many drugs, the effects after an overdose are related to the pharmacological actions of the drug, enhanced to the extent that they become toxic. Such reinforced pharmacological effects occur, for example, after overdose with hypnotics, sedatives, neuroleptics, antidepressants, opiates, central nervous system stimulants, beta-receptor blocking agents, calcium antagonists, belladonna alkaloids, ergot alkaloids, antidiabetic drugs and anticoagulants.

Many toxic effects, especially those caused by drugs, may be explained by excessive stimulation or blockade of receptors in cell membranes or within the cell. Many of the examples given above refer to interference with receptors in the central and peripheral nervous systems and in the cardiovascular system. Ion channels, which allow a selective influx or efflux of ions through cell membranes, may also be influenced by xenobiotics and, if this becomes excessive, severe toxicity may ensue. Poisoning by calcium-channel blockers is one of the most complicated poisonings seen in clinical toxicology.

Poisons may also disturb cellular metabolism through inhibition of enzymes. This is the mechanism in cyanide poisoning, in which cytochrome oxidase acitivity is blocked. Amatoxins from *Amanita phalloides* and related mushrooms block mRNA polymerase II, resulting in multiple clinical effects. Toxic substances, e.g. metals, may also block enzyme activity by reacting with catalytically essential SH-groups of enzymes.

Toxins may themselves be enzymes. For instance, snake venoms contain numerous enzymes that exert a wide spectrum of effects. The mixture of enzymes differs between species, so that toxic symptoms may be very variable. Venom enzymes may cause local tissue damage, coagulation disorders, neuromuscular block and circulatory instability related to release of endogenous, pharmacologically active agents (e.g. histamine, bradykinin, prostaglandins).

Poisoning may result in functional, mostly transient, disturbances in biological systems, or it may cause a chemical injury to bodily organs. Among drugs, paracetamol (causing liver damage) is an important example of the latter, as are methanol (retinal damage) and ethylene glycol (renal damage) among chemicals. These three agents also illustrate that, while the parent substance is relatively harmless, poisoning is instead related to highly toxic metabolites. Other examples of toxic lesions are renal damage caused, for example, by NSAIDs (nonsteroidal anti-inflammatory drugs), aminoglycosides, lithium or orellanine-containing mushrooms, and methaemoglobinaemia induced by oxidising agents.

Circumstances of drug poisoning

In Western countries, most severe drug intoxications are intentional, but accidental poisoning is also a significant problem.

Accidental poisonings

Poisoning incidents are common among *small children*, who are happy to pick up things and put them into the mouth. Most of these accidents occur in the age group 1–3 years and may involve all sorts of toxic agents, including medicines. Fortunately, the majority of these accidents result in no or only minor symptoms, largely owing to more efficient prophylactic measures and an increasing awareness of toxic risks.

Wrong dose, erroneous administration and confusion of medicines may cause poisoning. Such mistakes occur both at home and in clinics. Although not very frequent, these incidents may cause severe poisoning in both children and adults. Children seem to be especially prone to this type of mistake because of errors in calculating the appropriate dose and in dilution procedures. Among immigrant populations, language problems are a common reason for therapeutic mistakes.

Unintentional 'therapeutic' overdose is a common phenomenon, particularly in acute pain situations. Patients with lumbago, fractures, toothache, migraine, etc., may take analgesics too much and too often, hoping for pain relief. This is particularly frequent with paracetamol, as the lack of immediate side-effects means that the patient does not get any early warnings.

Intentional poisonings

Suicidal or parasuicidal intent is, by far, the most common cause of significant overdose of drugs in the Western world. Intentional drug poisoning occurs most frequently with psychotropic drugs (hypnotics, sedatives, antidepressants, neuroleptics, anticonvulsive agents, lithium); the second most commonly involved group is analgesics (e.g. paracetamol, NSAIDs, salicylates, opioids). In developing countries, pesticides, chemicals and natural toxins may dominate.

Substances of abuse may cause acute toxic symptoms in addition to their chronic toxicity. This may be due to an intentional overdose but is more often related to lack of knowledge about the actual dose, confusion of substances, impurities in the preparations and irregularities in the individual abuse pattern. In addition to specially designed drugs of abuse, regular medicines such as opioid analgesics and certain hypnotics and sedatives are also widely abused.

Criminal poisoning occurs but is less common. One important example, however, is the insidious form of child abuse in which medicines in overdose are deliberately given to small children. Generally one of the parents is responsible – a phenomenon known as Münchhausen by proxy. It is important to be aware of this possible aetiology in unclear cases, where poisoning cannot be ruled out. Another, increasingly common, type of criminal poisoning is the use of 'date rape drugs' administered in soft or alcoholic drinks. Substances involved are include flunitrazepam, gamma-hydroxybutyrate (GHB) or ethanol, and the purpose is mostly sexual abuse or robbery.

Prevention

Self-inflicted poisonings have a psychosocial background, and should be addressed accordingly. It can be argued, however, that withdrawal from the market of especially toxic drugs and replacement with less dangerous alternatives has proved beneficial to this group of patients in reducing morbidity and mortality after deliberate overdose. Examples are the introduction of new, less toxic sedatives, hypnotics and antidepressants.

A number of preventive measures have reduced both frequency and severity of accidental poisonings, particularly in children. Smaller packages for medicines, blister packs, child-resistant closures, adequate labelling, systematic information and education activities have all contributed to this favourable trend.

Management of drug overdose

Diagnosis

Adequate treatment of poisoning will require a proper diagnosis. The diagnostics aim at confirming the type and degree of poisoning. Diagnosis is based on history, observed clinical features and clinical investigations, e.g. laboratory analyses, radiography, CT scanning and endoscopy.

The *case history* is the most essential item and every effort should be made to clarify the toxic agent(s) involved, the probable dose and the time of exposure.

Clinical features may be characteristic and helpful in making the diagnosis (e.g. in poisoning with anticholinergics, central nervous system stimulants, digitalis, opioids, salicylates, theophylline). More often, however, the symptoms are, at least in the early stage, less specific and distinctive (e.g. in poisoning with antiepileptics, most cardiovascular drugs, hypnotics, sedatives, paracetamol).

Useful *laboratory analyses* are plasma glucose, haematology, acid–base and electrolyte balance, and creatinine and routine urine analysis. When appropriate, liver enzymes and liver function tests, muscle enzymes and coagulation tests can be added. It is especially crucial to detect and evaluate a severe metabolic acidosis – poisoning by methanol, ethylene glycol, salicylates and cyanide must be confirmed or excluded.

Toxicological analyses are useful in the acute stage, whenever the analytical results provide a guidance to specific or advanced treatment (certain antidotes or dialysis procedures). Drug poisonings, where a rapid toxicological analysis is essential for appropriate treatment, are given in Table 11.1.

In contrast, poisonings by benzodiazepines, anticholinergics and opioids, for example, present with typical symptoms, so that analytical confirmation is not needed to guide antidote treatment. A forensic toxicology screening may be useful in unclear, serious cases, and whenever there is suspicion of a crime.

Treatment

Treatment of poisoning includes four main principles: (1) decontamination, (2) symptomatic and supportive care, (3) enhanced poison elimination and (4) antidote therapy.

Decontamination

Only exposure by ingestion will be considered. Inhalation and skin contact are less relevant when discussing poisoning by drugs. Decontamination means that toxic agents not yet absorbed are removed or 'neutralised', e.g. through binding to activated charcoal. Decontamination must be done as early as possible.

Most poisonings occur after *ingestion*. Under certain circumstances, when time is the most crucial factor, there are possibilities to limit the absorption of poison. This can be done through administration of activated charcoal, gastric emptying and whole-bowel irrigation.

Activated charcoal binds most drugs effectively. Exceptions are small, charged molecules like iron and lithium. Efficacy is well documented for the first few hours after exposure. Charcoal is itself

Table 11.1 Drug analyses that provide guidance on special treatment, either with specific antidotes or advanced methods for poison elimination (e.g. haemodialysis)

Drug	Treatment
Carbamazepine	Elimination
Digitoxin, digoxin	Antidote
Iron	Antidote
Lithium	Elimination
Paracetamol (acetaminophen)	Antidote
Salicylate	Elimination
Theophylline	Elimination
Valproate	Elimination

non-toxic but may be harmful if it is aspirated into the lungs, which is a risk in unconscious patients.

The use of *gastric emptying* has declined over recent years, as studies have shown that efficacy diminishes rapidly with time. It is also an arduous and sometimes hazardous procedure. Generally it is undertaken if a severe poisoning is expected and the patient is admitted within 1–2 hours. The methods are either emesis induced by *ipecac syrup*, or *gastric lavage*. Ipecac syrup is nowadays rarely used and in some countries it has been entirely abandoned. Gastric lavage is the only method if the patient is drowsy or unconscious, and if used it must be preceded by endotracheal intubation to protect the airways.

Whole-bowel irrigation, using polyethylene glycol preparations, for example, can be considered if large amounts of highly toxic agents have been ingested, especially if these do not bind to charcoal. Examples are iron and lithium, especially in slow-release preparations.

Miscalculations and mix-ups may lead to *injection of toxic doses* of drugs in medical services. Furthermore, it is not uncommon among drug abusers that excessive doses are injected unintentionally. Other examples of toxic 'injections' are bites and stings by venomous animals. After injection of poisonous agents, the exposure cannot be reduced or stopped, and treatment must instead focus on treating the poisoning.

Symptomatic and supportive care

Most poisonings are managed by symptomatic and supportive care only. This means that symptoms occurring are treated as required. During a critical phase there may be a need for advanced support of vital functions: administration of intravenous fluids, correction of electrolytes and acid–base balance, inotropic and vasopressor support, antiarrhythmics, respiratory support and treatment of organ failure (e.g. hepatic, renal) are examples. In fact, the availability of good intensive care facilities is one major reason why the overall hospital mortality due to poisoning in developed countries is nowadays low (around 0.5%).

Enhanced poison elimination

Under certain circumstances it is possible to enhance the elimination of a poison, depending on the physiochemical properties and pharmacokinetics of the substance. Factors to be considered are volume of distribution, protein binding, molecular weight, metabolism, excretion and water solubility. For instance, a large volume of distribution and high protein binding in plasma will make it less likely that methods for enhanced elimination will be effective. Water solubility and molecular weight will also influence the possibilities of increased renal excretion and the efficacy of dialysis.

The use of multiple-dose activated charcoal is the most commonly applied method. Activated charcoal is given by mouth or through a gastric tube at 2–4 hour intervals during the acute phase of poisoning. The rationale for this is that charcoal may shorten the half-life of certain poisons by interrupting enterohepatic recirculation and through 'gastro-intestinal dialysis'. Drugs for which accelerated elimination has been documented using this method include carbamazepine, dapsone, dextro-propoxyphene, digitoxin and digoxin, phenobarbital, quinine and theophylline.

Alkaline diuresis will, by the principle of non-ionic diffusion, effectively promote renal excretion of weak acids, such as aspirin (acetyl-salicylic acid) and phenobarbital. Alkalisation is achieved through administration of bicarbonate intravenously and by the control of pH in blood and urine. Urine pH should be 7.5–8.0 to be effective.

Haemodialysis is effective and may be life-saving in severe poisonings by ethylene glycol, methanol, lithium and acetylsalicylic acid. Dialysis has also proved useful in a few other poisonings, but the indication is restricted to severe cases where symptomatic care proves insufficient. Examples are barbiturates, carbamazepine, meprobamate, sodium valproate and theophylline. If traditional haemodialysis is not available, a continuous method may be used instead. However, the continuous methods are slower in detoxifying the patient.

Antidote treatment

Antidotes are substances that in a specific way counteract the toxic effects of specific xenobiotics. In certain poisonings, antidote treatment is mandatory, such as those with paracetamol, iron, methaemoglobin-forming agents, digitalis, cholinergics, anticoagulants, methanol, ethylene glycol, cyanide, heavy metals and certain animal (e.g. snake) venoms. In contrast, some otherwise useful antidotes can, in the hospital setting, be replaced by symptomatic and supportive care, e.g. naloxone in opioid poisoning and flumazenil in benzodiazepine poisoning.

Antidotes exert their actions by widely varying mechanisms, as illustrated below. The principles of antidote action are displayed schematically in Table 11.2, where the most important antidotes and their 'target' poisons are mentioned.

Table 11.2 Overview of antidote mechanisms, important antidotes (in italic type) and their 'target poisons'

Mechanism	Antidotes and targets
Formation of inert complexes The poison is bound in a stable complex that can be excreted	• Chelating agents such as *desferrioxamine* (iron), *BAL, penicillamine, DMSA, DMPS, Ca-EDTA* (other heavy metals) • *Immunoglobulins* (snake, spider and fish venoms, digitalis, botulinus toxin) • *Hydroxocobalamin* (cyanide), *calcium* (fluorides)
Inhibited formation of toxic metabolites The antidote interferes with metabolism, preventing the formation of toxic metabolites	• *Ethanol* (methanol, ethylene glycol) • *Fomepizole* (methanol, ethylene glycol)
Enhanced endogenous detoxification capacity The antidote provides a substrate for strengthening endogenous detoxification	• *N-Acetylcysteine* (paracetamol) • *Sodium thiosulfate* (cyanide) • *Folinic acid* (methanol)
Interference at receptor sites Antidotes interfere with the poison at receptor sites, e.g. in the CNS or cardiovascular system	• *Naloxone* (opioids) • *Atropine* (cholinergics) • *Physostigmine* (anticholinergics) • *Flumazenil* (benzodiazepines and related agents) • *Blockers and stimulants of alpha- and beta-adrenergic receptors* (agents with the reverse effects on receptors)
Healing or limitation of biochemical or cellular injuries A heterogenous group including some well-documented and important antidotes and some less well-established ones	• *Vitamin K (phytomenadione)* (warfarin) • *Methylthionium chloride* (methaemoglobin inducers) • *Oximes* (organophosphorus pesticides, nerve gases) • *Biperiden, benzatropine* (neuroleptics) • *Pyridoxine* (isoniazid and gyromitrin from the mushroom *Gyromitra esculenta*) • *Silibinin* (amatoxins from mushrooms such as *Amanita phalloides* and *A. virosa*)

BAL, dimercaprol; DMSA, succimer (*meso*-2,3-dimercaptosuccinic acid); DPMS, unithiol (sodium 2,3-dimercaptopropanesulfonate); Ca-EDTA, sodium calciumedetate.

Formation of non-toxic complexes

The antidote binds to the toxic agent and forms a water-soluble complex that is inactive and can be eliminated through the kidneys. Examples are *chelating agents* used for detoxification in heavy-metal poisoning, and specific *immunoglobulins* that have been produced to neutralise snake venoms, botulinus toxin and digitalis glycosides. *Hydroxocobalamin* rapidly forms a stable complex with cyanide ions.

Inhibition of formation of toxic metabolites

In some poisonings the parent substance is relatively harmless compared to the metabolites. For a few of these it is possible to prevent the metabolism of the 'poison' by blocking the enzymes involved in metabolic degradation.

The two most prominent examples are poisoning by methanol and by ethylene glycol. Methanol poisoning will result in production of formic acid, causing metabolic acidosis and retinal damage. Ethylene glycol will form glycolic acid and oxalate, resulting in metabolic acidosis and renal damage.

The toxifying step is dependent on alcohol dehydrogenase (ADH). *Ethanol* has a much higher affinity for ADH than do methanol and ethylene glycol. At an ethanol concentration of around 22 mmol/L, oxidation of these chemicals by ADH is stopped: methanol and ethylene glycol are thus left untouched, generation of toxic metabolites will cease, and the chemicals are excreted unchanged.

While ethanol is a substrate for ADH, a recently introduced antidote, *fomepizole* (4-methylpyrazole), competitively blocks ADH, which also will inhibit metabolite formation (Figure 11.1). As fomepizole is lacking the undesirable effects of high-dose ethanol, it is the preferred antidote in severe cases.

Enhanced endogenous detoxification capacity

An ingenious antidote approach is to enhance the endogenous detoxification capacity of the body itself. There is one outstanding example in this category: the use of *N-acetylcysteine* (NAC) in paracetamol poisoning. Paracetamol is bioactivated to a strongly reactive intermediate metabolite (NAPQI; see Chapter 2). This metabolite is instantly bound to hepatic glutathione and excreted. After an overdose, however, the liver glutathione is depleted and the toxic metabolite will accumulate and bind to hepatocellular macromolecules, resulting in hepatic

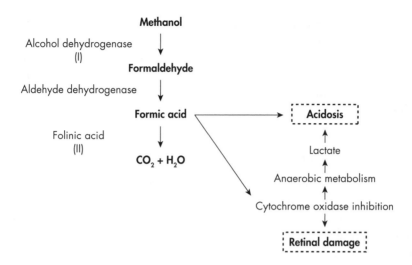

Figure 11.1 How the noxious effects of methanol can be eliminated by antidotes: (I) *ethanol* and *fomepizole* both inhibit methanol metabolism and the formation of toxic metabolites – this is the crucial treatment issue. (II) *Folinic acid* may enhance metabolism of formic acid – this effect on the clinical course is, however, far less important that that in (I).

necrosis. If NAC – a glutathione precursor – is administered within 8–10 hours after overdose, glutathione synthesis will be maintained and toxicity eliminated.

Another example is *sodium thiosulfate*, which, in cyanide poisoning, will replenish endogenous thiosulfate and enhance formation of non-toxic thiocyanate. Finally, in methanol poisoning where metabolic acidosis is already present, *folinic acid* may be beneficial in enhancing conversion of formic acid to carbon dioxide and water (see Figure 11.1).

Interference at receptor sites

Interference at receptor sites is perhaps the most classical and 'purest' antidote mechanism. It applies to true pharmacological receptor activity.

Naloxone may be life-saving in opioid overdose with severe respiratory depression, especially when emergency facilities are not immediately available. *Flumazenil* is effective in reversing CNS depression in poisoning by benzodiazepines and related drugs.

Another important antidote in this group is *atropine*, to be used in cholinergic over-stimulation caused by certain drugs, organophosphorus pesticides or muscarine-containing mushrooms. *Physostigmine*, a cholinesterase inhibitor that passes the blood–brain barrier, may be useful in reversing central anticholinergic effects caused by certain drugs or plants containing belladonna alkaloids. *Alpha-* and *beta-adrenergic receptor stimulants* or *blockers* may be used as antidotes, depending on whether blockers or stimulators have been overdosed.

Repairing or limiting biochemical or cellular injuries

This is a heterogeneous group of antidotes, the effects of some of which are well defined while they are less convincing for others.

Vitamin K (phytomenadione) is the logical and effective antidote in warfarin poisoning. *Methylthionium chloride* (methylene blue) effectively reduces methaemoglobin back to haemoglobin, restoring the oxygen transport capacity. Drugs such as prilocaine, dapsone and phenazone are methaemoglobin inducers.

Oximes (obidoxime or pralidoxime) are cholinesterase reactivators, used in poisoning by organophosphorus insecticides and nerve gases, where cholinesterase inhibition will cause life-threatening cholinergic and nicotinergic overstimulation. *Biperiden, benzatropine* and similar drugs may counteract extrapyramidal symptoms caused by neuroleptics.

Pyridoxine will restore the CNS content of pyridoxal phosphate that has been depleted by isoniazid or gyromitrin (occurring in the morel *Gyromitra esculenta*).

Special poisonings

Used properly, drugs act in a balanced way and to the patient's benefit. In overdose, they may turn into powerful poisons. Some commonly occurring drug poisonings, illustrating the main principles of poisoning treatment, are briefly presented below.

Benzodiaxepines and related drugs

The replacement of older, more toxic agents by benzodiazepines and related drugs has reduced morbidity and mortality related to poisoning in this group.

Main symptoms: sedation and unconsciousness. Respiratory depression and cardiovascular disturbances are unusual.

Treatment: mainly symptomatic. The benzodiazepine antagonist *flumazenil* may be useful in small children, in the elderly, and as a diagnostic tool. It is seldom needed in the routine management of healthy adults.

Tricyclic antidepressants

The use of tricyclic antidepressants has declined over the last decade, and this has reduced the overall mortality and morbidity from antidepressant poisoning.

Main symptom: CNS depression, seizures and serious cardiovascular disturbances.

Treatment: symptomatic and supportive.

Selective serotonin reuptake inhibitors

Selective serotonin reuptake inhibitors (SSRIs) nowadays dominate the market for antidepressants. Because of their lower toxicity compared to tricyclic antidepressants, this is a positive trend.

Main symptoms: CNS depression and seizures; severe cardiovascular disturbances are uncommon.

Treatment: symptomatic and supportive.

Neuroleptics

Neuroleptics are commonly overdosed.

Main symptoms: CNS depression and extrapyramidal symptoms, seizures (in heavy overdose). Cardiovascular symptoms are less pronounced.

Treatment: symptomatic and supportive. Extrapyramidal symptoms may be counteracted by *biperiden* or *benzatropine*, for example.

Lithium

Gradual onset of renal dysfunction, starvation, fluid loss or infection may cause an accumulation of lithium, resulting in toxic levels and symptoms of chronic poisoning; however, acute poisonings also occur.

Main symptoms: neurological and renal.

Treatment: enhancement of lithium elimination as quickly as possible; in cases of persisting high concentrations, kidney dysfunction or serious symptoms, haemodialysis should be applied. Symptomatic and supportive care.

Clinical toxicology

Anticholinergics

A number of drugs have both peripheral and central anticholinergic properties, but most typical are the belladonna alkaloids and their derivatives.

Main symptoms: tachycardia, mydriasis, fever, confusion, hallucinations, seizures and possibly coma.

Treatment: symptomatic and supportive. Under certain circumstances, *physostigmine* is indicated and will reverse most symptoms.

Aspirin (acetylsalicylic acid)

Poisoning by salicylates is nowadays less common.

Main symptoms: gastrointestinal discomfort and 'salicylism' (vertigo, tinnitus, impaired hearing, anxiety, sweating, hyperventilation and fever); later, metabolic acidosis, hyperthermia and convulsions.

Treatment: symptomatic and supportive. In significant poisoning, measures to enhance elimination through alkaline diuresis or haemodialysis are mandatory.

Paracetamol (acetaminophen)

If properly used, paracetamol has remarkably few side-effects. In contrast, overdose means a well-defined risk of acute hepatic damage. The problem with subacute, 'therapeutic' overdose has been discussed above (see p. 196).

The mechanism of paracetamol toxicity and the role of *N*-acetylcysteine as an antidote are discussed in Chapter 2 and above on p. 202.

Main symptoms: occasionally, initial gastric discomfort; after 2–3 days, signs of liver damage, possibly resulting in hepatic failure.

Treatment: *N-acetylcysteine*, if given within 8–10 hours, will prevent liver damage.

Opioids

Many drugs are found in this category and the spectrum of pharmacological effects is wide.

Main symptoms: CNS depression, pronounced respiratory depression and maximum miosis. Poor correlation between respiratory depression and level of consciousness. Bradycardia, hypotension.

Treatment: symptomatic and supportive. *Naloxone* is effective in reversing opiate-induced respiratory depression.

Digitalis

Chronic poisoning may develop during treatment because of reduced metabolism, impaired excretion, drug interactions, dehydration, etc.

Main symptoms: nausea and vomiting, fatigue, confusion, disturbed colour perception, arrhythmias.

Acute massive overdose occurs, but more rarely.

Main symptoms: nausea and vomiting, arrhythmias, hyperkalaemia, cardiac failure, CNS depression.

Treatment: symptomatic and supportive. In severe poisoning, *digitalis-specific antibodies* are effective and may prove life-saving.

Iron

Because of its medical use, iron is widely available and represents the most common metal poisoning. However, because of effective preventive measures, severe iron poisoning as a childhood accident is nowadays rare.

Main symptoms: abdominal pain, vomiting, diarrhoea, CNS depression, metabolic acidosis, circulatory instability and hepatic injury.

Treatment: decontamination is important and may, in addition to gastric lavage, include whole-bowel irrigation. *Desferrioxamine* chelates iron and facilitates its excretion. Symptomatic and supportive care.

Conclusion

Acute poisoning is a common medical emergency worldwide. Poisonings may be accidental or deliberate. Most severe poisonings are self-inflicted, and in Western countries drugs and substances of abuse dominate heavily. Elsewhere, chemicals and pesticides may be the main problem. Among the many routine cases, the exceptions to the rule may appear quite unexpectedly: unclear cases presenting diagnostic challenges, odd cases, and those with an exceptional toxicity where extreme treatment procedures are required. The complexity is considerable within clinical toxicology when it comes to circumstances of poisoning, the wide variety of possible toxic exposures, and the wide range of therapeutic options. Management of poisoning requires precision and promptness in diagnosis and treatment.

Further reading

Brent J, *et al.* (2005). *Critical Care Toxicology – Diagnosis and Management of Critically Poisoned Patients*. Philadelphia: Elsevier Mosby.

Flanagan RJ, Jones AL (2001). *Antidotes*. London: Taylor & Francis.

Goldfrank LR, *et al.* (2002). *Toxicological Emergencies*, 7th edn. New York: McGraw-Hill.

Jones AL, Dargan PI. (2001). *Churchill's Pocketbook on Toxicology*. Edinburgh: Churchill Livingstone.

12

Safety assessment of pharmaceuticals: regulatory aspects

Jan Willem van der Laan

European legislation in a global environment

Regulation of medicines in Europe started shortly after the thalidomide affair (see Chapter 4). This toxicological disaster led in many European countries to rapid establishment of so-called 'competent authorities', which were responsible for admission of drugs to the market as well as drug surveillance. Europe was several decades behind the USA, where the Food and Drug Administration (FDA) had already started shortly after 1900. The purpose of the legislation by the European Union (EU) was to establish a common market essentially without borders for human pharmaceuticals. This economic background is still the driving force of the administration in Europe, having the power of legislation under the Directorate General (DG) 'Enterprise' and not under the DG 'Sanco' (Public Health and Consumer Products). In 1975 the European Commission described harmonised views regarding the analytical, pharmacotoxicological and clinical standards and protocols in respect of testing of proprietary medicinal products. Vaccines, serums, allergens, immunological medicinal products, radiopharmaceuticals and medicinal products derived from human blood or human plasma were included later.

Under the new European Commission, all the directives have been brought together under the new Directive 2001/83/EC, with an Annex 1 defining the Analytical, Pharmacotoxicological and Clinical Standards and Protocols in Respect to the Testing of Medicinal Products.

In 1995 the European Agency for the Evaluation of Medicinal products (EMEA), currently called the European Medicines Agency (but still abbreviated as EMEA), was established in London to stimulate a harmonised interpretation of rules governing medicinal products by the different EU member countries. The Committee for Medicinal Products for Human Use (CHMP) is responsible for advice with regard to the marketing authorisation of new human pharmaceuticals, which are subject to centralised Community Authorisation by the European Commission.

Various CHMP working parties consisting of expert members write guidelines that give guidance to the industry in developing new drugs. The main working parties are:

- Efficacy Working Party, dealing with clinical efficacy of medicinal products
- Pharmacovigilance Working Party, dealing with the clinical safety of products that are on the market
- Safety Working Party, dealing with the experimental pharmacology and toxicology
- Quality Working Party, dealing with the chemical manufacturing concerns with regard to conventional chemical medicinal products
- Biologics Working Party (formerly Biotechnology Working Party) dealing with manufacturing concerns regarding biological products (except herbal medicinal products)

Additional working parties exist with regard to the use of medicinal products in children (Paediatric Working Party) and on Vaccines (Vaccines Working Party), among others.

Since 1990 there has been an initiative of the International Conference of Harmonisation for Technical Requirements on Pharmaceuticals for Human Use (abbreviated as ICH). The partners consist of three regulatory parties (the FDA from the USA, the Ministry of Health, Labour and Welfare from Japan, and the European Commission with the EMEA from Europe), as well as three industrial partners (the US Pharmaceutical Research Manufacturers Association (PhRMA), the Japanese Pharmaceutical Manufacturers Association (JPMA), and the European Federation of Pharmaceutical Industry Associations (EFPIA)).

The aim of the ICH is to harmonise the requirements on human medicinal products over the main markets in the world. Since the start of the ICH, the essential elements in the development of pharmaceuticals have been described in guidance documents, which are incorporated in the legislation of the various areas (see Table 12.1).

Good Laboratory Practice

In the second half of the 20th century, several cases of fraud in the USA prompted the authorities to devise regulations to prevent such

Table 12.1 Safety topics of the International Conference on Harmonisation

Number	Topic
S1	Carcinogenicity
S2	Genotoxicity
S3	Toxicokinetics
S4	Duration of the repeated dose toxicity study, rodents and non-rodents
S5	Reproductive toxicity
S6	Safety of biotechnology products
S7	Safety pharmacology
S8	Immunotoxicity
M3	Nonclinical studies needed for clinical trials
M4	Common Technical Dossier

problems. A system of Good Laboratory Practice (GLP) was developed in the 1970s to ensure quality. The use of GLP methodology does not guarantee good scientific quality per se: a critical assessment of the methodology is still needed when all studies are carried out under GLP regulation. GLP principles applied in Europe are described in the OECD (Organisation for Economic Co-operation and Development) Series on Principles of Good Laboratory Practice and Compliance Monitoring.

All non-clinical safety studies used for marketing authorisation of pharmaceuticals in Europe should meet GLP standards. This includes safety pharmacology studies and all toxicity studies as well as the toxicokinetic studies carried out to support the toxicity studies. Developmental pharmacological studies do not come under the GLP regime.

The Clinical Trial Directive

Regulatory approval of clinical trials differs greatly in various countries around the world. The FDA has implemented its Investigational New Drug (IND) process: for all clinical trials, starting with the first Phase I studies, companies have to submit a dossier to the FDA in order to obtain permission to start. To avoid delay, response from the FDA is expected within 60 days. An FDA division can halt the progress of the study ('put on hold') within 30 days after receipt of the documentation. Often companies go to the FDA for a pre-IND meeting during the first developmental phase, to discuss the package needed for the first use in humans.

Until recently, the situation in Europe was very diverse, with some countries having approval procedures comparable to that of the FDA, others having a very different approach. In 2004 the new European Clinical Trial Directive 2001/20/EC was implemented in most member countries, harmonising the requirements in the EU. The Declaration of Helsinki (see Table 12.2) is a strong statement of the need for careful safety studies carried out in animals. The Investigators Brochure should comply with the ICH guideline on Good Clinical Practice. The documentation of the company is described in the Investigational Medicinal Product Dossier, comparable to the IND dossier of the FDA. Directive 2001/20/EC applies to all investigational medicinal products, including biotechnology products, immunological medicinal products and herbal medicinal products.

Marketing applications: format and procedures

Companies have to submit extensive documentation for their marketing applications. The ICH process has defined a Common Technical Dossier (CTD), which precisely defines the structure of the documentation (see Figure 12.1). The CTD is accepted by the EU as the only way in which documentation can be applied.

The CTD only defines the place in the dossier where the documentation has to be included. In no way should the CTD be interpreted as a list of studies that are needed. For example, mention of the item 'juvenile toxicity studies' does not mean that these studies are always needed. It serves only to indicate that if such studies are needed, they should be stored under the relevant number.

Table 12.2 Declaration of Helsinki on the need for toxicity studies

§1	Biomedical research involving human subjects must conform to generally accepted scientific principles and should be based on adequately performed laboratory and animal experimentation and on a thorough knowledge of the scientific literature.
§5	Every biomedical research project involving human subjects should be preceded by careful assessment of predictable risks in comparison with foreseeable benefits to the subjects or to others. Concern for the interests of the subject must always prevail over the interests of science and society.

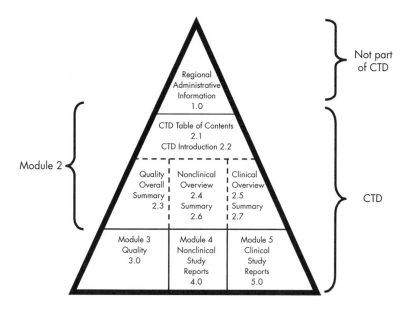

Figure 12.1 The various modules of the Common Technical Dossier. From EMEA: http://www.emea.eu.int/pdfs/human/ich/288799enm.pdf.

Assessment of safety is always carried out in relation to therapeutic efficacy

Pharmaceutical companies are responsible for the safety of their products. The concept of safety can only be examined in relation to therapeutic efficacy: the 'benefit/risks' assessment. The final assessment is done at the stage of the marketing authorisation by the competent authority, but it can be seen as a continuous process during the development of a product by the company as well as by the regulatory authorities in the approval for clinical studies.

An important issue is the impact on the health and illness of the patients.

Examples

• *Vaccine* If the product is a vaccine it will be given to healthy people, in most cases children, as a preventive method. Only minor side-effects are acceptable. For example, much resistance exists in the UK and Sweden against vaccination with whole-cell pertussis vaccine because of serious side-effects (convulsions, persistent screaming).

- *Cytostatic anticancer product* If the product is a cytostatic compound intended to treat patients with a short life-expectancy, the acceptance of adverse effects is much higher; even complete hair loss and a high risk of bacterial infections (as the result of an immunosuppressive effect) would be acceptable.

  Cytostatic anticancer drugs have an intrinsic risk of genotoxicity and this is openly accepted in a guideline on this type of product, whereas for other products developed for non-life-threatening indications, genotoxicity is not acceptable.

- *Sleeping pill* A strong effect on blood pressure, for example, might be acceptable for a life-saving drug, but it would represent a high risk for patients taking it as a sleeping pill each night.

Competent authorities, industrial companies, and also prescribers and patients should be aware that no pharmaceutical compound is without risk, and this risk should be carefully weighed against the benefit of the compound.

In the development of a new drug, this relation between benefit and risk becomes operational with comparison of the pharmacologically active dose and the toxicological dose. Establishing a pharmacologically active dose is not always an easy task. In experiments *in vitro*, a dose–response study can be done and an ED_{50} value determined; however, such detailed results are not always easy to obtain in experiments *in vivo*: choice has to be made of which end-point is most related to the intended therapeutic action in humans. The same is true for the toxicological dose. The choice of end-point is important: is it the No-Observed-Adverse-Effect-Level (NOAEL) or simply the No-Observed-Effect-Level (NOEL)? This demands judgement on what constitutes an adverse effect. Furthermore, many toxicological effects of human pharmaceuticals are related to their pharmacological effects, and can be seen as exaggerated effects. It can therefore be difficult to calculate a ratio between the pharmacologically active dose and the NOAEL (if it exists).

Another possibility is that the toxic symptoms appearing at a low dose (in relation to the pharmacologically active dose) are acceptable in a clinical sense. In this case another end-point has to be chosen, mostly at a higher dose, in another organ. Because of these complications, it may be difficult in some cases to establish a benefit–risk ratio.

Assessment of safety requires extrapolation from animal studies to human effects and exposure

Two aspects are important in assessing the risks of exposure to human pharmaceuticals for humans on the basis of animal toxicity:

* Comparison of effects
* Comparison of exposure

For small molecules (conventional pharmaceuticals), the comparison of *effects* is usually not a major issue, as the active site 'receptors' are similar in various species and the affinities in human tissues are comparable to those in animal tissues. Species differences may be much more important in the case of biotechnology-derived products, such as recombinant proteins and monoclonal antibodies (see below).

The comparison of *exposure* is a more difficult issue, as the pharmacokinetic profiles differ much more between animals and humans, especially in the case of small molecules. For an initial assessment to estimate the first dose in humans, the researcher has to rely on allometric methods, taking into account the differences in metabolism between various species. Allometric scaling is a procedure based on differences in metabolism between animal species; the comparison is done from experience on the basis of body surface and not on body weight. This basis derives from the comparison of metabolism rates of individual organs, but ultimately the comparison on the basis of body surface is sufficient in practice.

For 30% of compounds this estimate will fit very well, but for the remaining cases it may be higher or lower, depending on factors including the activity of the various cytochrome P450 or Phase 2 isoenzymes, (see Chapter 2). For risk assessment during the process of drug development, the preferred method is calculation of the ratio between human and animal exposures on the basis of the AUC (area under the curve) values. The Margin of Safety is then defined as the ratio between the AUC at the NOAEL (for the NOAEL see below) of the relevant animal and the AUC in humans at the intended therapeutic dose (also keeping C_{max} (maximum concentration) differences in mind since these might result in differences in acute toxicity).

Safety assessment is a stepwise process

Safety assessment is an iterative process carried out during development in close cooperation between pharmacologists, toxicologists and clinicians. The developmental process is conducted in phases, before the first use in humans as well as afterwards. The first step in research is to think about the direction along which a new compound will develop: a leading hypothesis is important to focus thinking about the possible (side-) effects of a new class of compounds. Selection of lead compounds follows after the stage of drug discovery. No general guidelines exist at this stage, and each company has established its own policy in its research and development (R&D) departments. An important issue already at this stage is high predictivity in the test results so as to have the lowest attrition rate possible. The predictivity of toxicity tests in animals will be discussed later.

On coming to the first use in humans (Phase I) the ICH 'Note for Guidance on Non-Clinical Safety Studies for the Conduct of Human Clinical Trials for Pharmaceuticals' (Topic M3) is applicable. The guideline refers to the different clinical phases of the development of human pharmaceuticals as defined in the ICH guideline 'General Considerations of Clinical trials' (Table 12.3).

Animal studies that should be conducted before the first use in humans are called more specifically *pre*-clinical studies. In ICH guidelines the term '*non*-clinical safety' studies is introduced in order to indicate that various animal safety studies can be conducted in parallel with clinical studies and are not necessary before the first

Table 12.3

Phase	Timing and action
Phase I	First human exposure: generally single-dose studies followed by dose escalation and short-term repeated dose studies to evaluate pharmacokinetic parameters and tolerance. Often conducted in human volunteers, but involvement of patients is possible In short: *human pharmacology studies*
Phase II	Exploratory efficacy and safety studies in patients In short: *therapeutic exploratory studies*
Phase III	Confirmatory clinical trials for efficacy and safety in patient populations In short: *therapeutic confirmatory studies*
Marketing authorisation	This can be done based on the outcome of the confirmatory clinical trials

time a compound goes into the clinic (the strict interpretation of *pre-clinical*).

The objectives of the animal safety studies are to define the pharmacological and toxicological effects of a compound, first in relation to each other, and later in relation to the effects in humans.

The first step in development of a new drug is its pharmacology, essentially the lead properties of the compound to be used to treat a targeted disease or specific symptoms. It is important to have data available as a type of 'proof-of-concept' – Why should this product be of help in the intended patient population?

Examples

- *Parkinson's disease*   The relation between dopaminergic properties of anti-parkinsonian drugs and their efficacy in the intended patient population is well-established. There are several subclasses of dopaminergic receptors present in animals. Companies work to select compounds with the relevant selectivity and specificity for this type of receptor in order to enhance the therapeutic efficacy.

  Even in this case the company is expected to show a pharmacological effect of the compound in appropriate animal models, i.e. in the so-called turning model, in which one side of the dopaminergic neurons in the corpus striatum is damaged.

- *Crohn's disease*   The efficacy of drugs is very difficult to predict from animal studies. The intestinal immune system is thought to be affected in this disease, but adequate animal models are lacking. New drugs in this area are developed mainly on the basis of mechanistic information gathered in human studies, e.g. by interactions with interleukins. Proof-of-concept is therefore based on very specific pharmacological properties, which are not necessarily tested in an animal model.

Using the proof-of-concept data, the potency of a compound can be derived. The establishment of a pharmacologically active dose is important as a comparator for the interpretation of the toxicity studies and the risk assessment. Toxicity studies with pharmacologically active compounds might be difficult to interpret as the pharmacological activity often has toxicological consequences. It is important, therefore, to think about possible explanations of toxicological effects as the result of pharmacological effects. Definition of the No-Adverse-Effect-Level (NOAEL) to estimate the safety margin may therefore sometimes be difficult.

Safety studies required to support first use in humans

Phase I

The safety assessment is the next step. The ICH M3 Guideline specifies that before the first use in humans the following data should be present:

- Safety pharmacology with respect to vital functions, cardiovascular, central nervous system and respiratory systems.
- Information on absorption, distribution, metabolism and animal metabolic pathways.
- Single-dose (acute) toxicity in two mammalian species. A dose-escalation study is considered an acceptable alternative.
- Repeated-dose toxicity. The duration is depended on the duration of the trial. As most of the Phase I studies will be carried out as single-dose studies, a duration of 2 weeks is sufficient. Studies should be done both in rodents and in non-rodents.
- Local tolerance should be studied, preferably in connection with the other toxicity studies
- Genotoxicity data from *in vitro* tests: evaluation of mutations (bacterial mutagenicity, Ames test) and of chromosomal damage (e.g. mouse lymphoma) is generally needed.

Discussions are ongoing whether this package is needed for compounds that will be given in humans only as a single low dose (e.g. an imaging agent, or a muscle relaxant during surgery). Companies are developing new types of strategies to obtain early data on pharmacokinetics in humans, e.g. with respect to bioavailability and metabolism.

The Safety Working Party has produced a 'Position Paper on Non-clinical Studies to Support Clinical Studies with a Single Microdose' (CPMP/SWP/2599/02/Rev1) indicating that the authorities will give the industry the possibility to start alternative approaches. The term 'microdose' is defined as 1/100 of the dose required to yield the primary pharmacological effect of the test substance (typically doses at or below the microgram range).

The main part of the reduction of the package is the introduction of 'extended single dose' studies in which the company is expected to include also the safety pharmacology, such as cardiovascular end-points. This study is meant to combine the safety pharmacology and the acute toxicity study, as well as a repeated-dose study. With low-toxicity compounds, a limit dose might be set at 1000 times the allometric dose equivalent from the animal to human dose. The observation period should be 14 days with an interim sacrifice at day 2 to detect acute effects.

The FDA has developed a specific procedure called 'Screening IND' (IND = investigational new drug). This procedure specifies the requirements for compounds that may even be tested together in the same volunteer, for example to produce a rapid comparison of the pharmacokinetics in the same person.

Further negotiations are underway to reduce the number of studies to be used for exploratory studies. The industry indicates that the attrition rate of compounds is highest in the first round after introduction in humans. However, some elements in the preclinical package mentioned above require a huge amount of the compound, especially the study in dogs to be carried out at the Maximum Tolerable Dose. This is a problem because the compound will be available only in limited quantities at that stage. A study in humans focused on pharmacokinetic characterisation may be based on a lower dose, and the safety of this lower dose can be studied using a fixed maximum dose. There are two possible positive points to this initiative:

- The requirement for the compound is decreased, which is important for the company as the classical approach often needs a scaling-up of the synthesis of the compound, which requires time and money.
- The selection of a lead compound from a group of several candidates can be done using a lower number of volunteers, exposed to a lower level of the compound. This is also advantageous from an ethical point of view.

Safety studies required to support further clinical trials and marketing authorisation

Phase II

Based on the regulations thus far, industry should supply data from 2-week studies in rodents as well as non-rodents to support Phase II trials of human pharmaceuticals that are intended to be used short-term, i.e. for less than 2 weeks. In case of further development, additional studies lasting at least 1 month are needed for marketing the drug (see below).

The requirements for Phase II 'Human Therapeutic Studies' are an extension of the studies mentioned (see Table 12.4).

Table.12.4 Duration of repeated-dose toxicity studies to support Phase I and II trials in the EU and Phase I, II and III trials in the USA and Japan.[a] Taken from the ICH M3 Guideline

Duration of clinical trials	Minimum duration of repeated-dose toxicity studies	
	Rodents	Non-rodents
Single dose	2 weeks[b]	2 weeks
Up to 2 weeks	2 weeks[b]	2 weeks
Up to 1 month	1 month	1 month
Up to 3 months	3 months	3 months
Up to 6 months	6 months	6 months[c]
>6 months	6 months	Chronic[c]

[a] In Japan, if there are no Phase II clinical trials of equivalent duration to the planned phase III trials, conduction of longer-duration toxicity studies is recommended.

[b] In the USA, as an alternative to 2-week studies, single-dose toxicity studies with extended examinations can support single-dose human trials.

[c] Data from 6 months of administration in non-rodents should be available before the initiation of clinical trials longer than 3 months. Alternatively, if applicable, data from a 9-month non-rodent study should be available before the treatment duration exceeds that which is supported by the available toxicity studies.

The standard battery of tests for genotoxicity, usually the *in vivo* test, should be completed prior to the initiation of Phase II studies. Needless to say, the level of exposure should be shown to be adequate in blood in order to validate the outcome of the study.

Phase III

Table 12.5 shows the requirements for a Phase III study; there are some regional differences as indicated in the ICH M3 Guideline.

Specific populations

Women of child-bearing potential: reproductive toxicity

There are regional differences with respect to the inclusion of women of child-bearing potential in clinical trials. In the USA equal rights legislation, the basis is that it is illegal to exclude women from clinical trials; accordingly, adequate contraception is important. As it is not always clear whether there might be an interaction with hormonal contraception, an additional barrier method is recommended to ensure that conception cannot occur in the presence of a new pharmaceutical

Table 12.5 Duration of repeated-dose toxicity studies to support Phase III trials in the EU and marketing in all regions.[a] Taken from the ICH M3 Guideline

Duration of clinical trials	Minimum duration of repeated-dose toxicity studies	
	Rodents	Non-rodents
Up to 2 weeks	1 month	1 month
Up to 1 month	3 months	3 months
Up to 3 months	6 months	3 months
>3 months	6 months	Chronic[b]

[a] This table also reflects the marketing recommendations in the three regions except that a chronic non-rodent study is recommended for clinical use >1 month.
[b] Data from 6 months of administration in non-rodents should be available before the initiation of clinical trials longer than 3 months. Alternatively, if applicable, data from a 9-month non-rodent study should be available before the treatment duration exceeds that which is supported by the available toxicity studies.

compound for which no data from reproductive toxicity studies are available.

In the EU, the authorities share the opinion that women of child-bearing potential can be included only after data on reproductive toxicity are available. They are very reluctant to allow studies in pregnant women unless a great deal of experience is available in non-pregnant women, in combination with animal data on reproductive toxicity (see Table 12.6). All available data on pharmacology and (repeated dose) toxicology should be taken into account in the assessment of the reproductive risks to humans.

It is recommended that the same species and strain be used as in toxicological studies. Rats and mice are the default rodent species, while the rabbit is a non-rodent species for which there is a lot of experience. If rabbits are unsuitable (e.g. in the case of biotechnology-derived species, see below), dogs or monkeys come into the picture, but in these cases there are statistical issues – for example, the monkey litter is much smaller.

Children: juvenile toxicity

The last 5–10 years have seen a lot of attention from policy makers to the use of pharmaceuticals in children. Many drugs used regularly in children have never been tested systematically in children, in keeping with the idea that children are just small adults and the dose can be adjusted simply by using the body weight ratio. However, important

Table 12.6 The stages of development agreed upon. Unless shown otherwise it is assumed for rats, mice and rabbits that implantation occurs on day 6–7 of pregnancy and closure of the hard palate on day 15–18 of pregnancy. Taken from ICH 5a

A	Premating to conception (adult male and female reproductive functions, development and maturation of gametes, mating behaviour, fertilisation
B	Conception to implantation (adult female reproductive functions, preimplantation development, implantation)
C	Implantation to closure of the hard palate (adult female reproductive functions, embryonic development, major organ function)
D	Closure of the hard palate to the end of pregnancy (adult female reproductive functions, fetal development and growth, organ development and growth)
E	Birth to weaning (adult female reproductive functions, neonatal adaptation to extrauterine life, preweaning development and growth)
F	Weaning to sexual maturity (postweaning development and growth, adaptation to independent life, attainment of full sexual function)

differences may exist that in the past have led to fatal toxicity. Because of these and other differences, there is active debate on the need for studies in juvenile animals in order to identify risks of this type.

Examples

- Valproate is an antiepileptic used especially, although not exclusively, for petit-mal epilepsy. Hepatotoxicity has been observed in children treated with valproic acid because of their higher vulnerability.
- Paracetamol (acetaminophen) is less toxic in young children because they have a greater capacity to metabolise paracetamol as a consequence of more active sulfation and a higher rate of glutathione turnover.

Specific products

There is general consensus that certain product categories in the field of human pharmaceuticals can be handled differently from conventional medicines – biotechnological products such as recombinant proteins and monoclonal antibodies, and also vaccines based on infectious material or even tumour cells.

Biotechnology

The guidance document ICH S6 *Preclinical Safety Assessment of Biotechnology-Derived Pharmaceuticals* expresses a common philosophy of testing in the various regions of the world:

> All regions have adopted a flexible, case-by-case, science-based approach to preclinical safety evaluation needed to support clinical development and marketing authorisation.

The physiological character of the compounds raises questions about the need for non-clinical evaluation, as many of the products are peptide/protein hormones that belong to the body. However, their use as therapeutics is associated with a non-physiological route of administration (mostly intravenous or subcutaneous) at concentrations that might be higher than will be ever be reached under normal physiological conditions. Thus, on the one hand, the compounds are well-characterised; on the other, some risks are unknown.

A further issue is the species-specificity of the compounds. Proteins from different sources, although similar in function, may be slightly different in amino-acid sequence, and are not always recognised with the same affinity across the various species. Moreover, post-translational modifications (e.g. glycosylation patterns) may be different between species. The species specificity and/or tissue specificity hampers standard toxicity testing in commonly used species such as rats and dogs. Safety studies in non-responsive species, i.e. species not responding to the pharmacological effect of a compound, are useless, as toxicological hazards cannot be compared with the pharmacological benefits in the same species, and no margin of safety can be estimated.

Examples

- Human insulin can be tested reasonably in rodents despite small differences in amino acid sequence.
- The von Willebrand factor in blood (factor VIII), however, is more species-specific and does not behave in a similar way in rodents; it can be tested only in non-rodents such as dogs and monkeys.

The ICH S6 guidance document therefore describes the general principles of testing of biotechnology-derived products to define pharmacological and toxicological effects.

Biological activity can be characterised using *in vitro* assays, and the use of cell lines and primary cell cultures can be helpful in this respect, for example to study effects on cell proliferation. Other aspects involve pharmacological end-points such as receptor affinity and occupancy, which might play a role in the selection of an appropriate animal species.

For monoclonal antibodies, the immunological properties should be described in detail, especially their antigenic specificity and complement binding. Species specificity is high for many monoclonal antibodies developed in recent years, making it very difficult to test their safety in an animal species other than monkeys, in some cases only in chimpanzees. Transgenic mice carrying the appropriate antigen are suggested as an alternative testing species for some aspects, but the uncertainty of the predictivity of such models is naturally high.

Of special importance is the possibility of unintentional reactivity and/or cytotoxicity towards human tissues distinct from the intended target. Cross-reactivity studies are needed using a range of human tissues. However, the occurrence of cross-reactivity *in vitro* does not always indicate a real risk *in vivo*, as some cross-reacting epitopes might be located intracellularly, not being reached by intravenous administration of the compound.

Because of the higher species specificity as well as the more physiological character of these compounds, there is no longer a need for studying toxicity in two animal species. A toxicological study in one relevant species may suffice in many cases; and even when two species might be necessary to characterise toxicity in short-term studies, the use of one species for long-term studies can be justified.

A special issue is the immunogenicity of biotechnology-derived products in animals. Since the compounds are designed to be as near-identical as possible with human proteins, there might be small differences from their animal counterparts, leading to recognition as non-self in animal species used for toxicity testing. It is important to be aware of the possibility that the induction of an antibody response might lead to differences in the pharmacokinetics of the compound (either increasing or decreasing its half-life) or even neutralisation of its activity. Accordingly, characterisation of antibody responses is needed to interpret the outcome of the study.

The possible immunogenicity, however, is in no way predictive of the potential to induce antibody formation in humans or of the risks associated with this induction.

Vaccines

Protection against several diseases can be obtained by administering vaccines to healthy people. Vaccines are preparations containing antigenic substances. These substances induce a specific and active immunity against an epitope of the infecting agent, and in that way a protection against the infectious agent itself.

Vaccines fall into various classes:

- Inactivated organisms
- Living organisms
- Antigens extracted from organisms, secreted by them or produced in some other way.

Inactivation may be achieved through detoxification by chemical or physical means. Antigens may be conjugated to a carrier to enhance their immunogenicity. Another possibility for increasing the effectiveness of a vaccine is the addition of an adjuvant.

Testing of the pharmacology and toxicology of vaccines needs a specific approach. The infectious disease for which an animal model has to be found is in many cases specific to humans in its clinical features, and the infectious agent may not induce symptoms of disease in animal species in the same way. Therefore, selection of an animal species should be made on a case-by-case basis. Enhancement of the intended disease, induction of local toxicity, and adverse immunological effects such autoimmunity or sensitisation are all potential safety concerns for vaccines.

Several guidelines on pharmacological and toxicological testing are available. The World Health Organization has recently developed an updated broad Guideline on Preclinical Testing of Vaccines (November 2004), which also discusses the toxicological requirements for the first use in humans.

The efficacy of vaccines is often enhanced by the use of adjuvants as mentioned above. Aluminium salts or the hydroxide have long been the only adjuvants approved for human purposes. A few years ago the discovery of Toll-like receptors (TLRs) gave a huge stimulus to the development of new adjuvants, which previously had been developed essentially on a trial-and-error basis. Several adjuvants appear to be specifically active on these receptors, such as mucopolysaccharide (MPL) on the TLR4 and oligonucleotide sequences such as CpG motifs on the TLR9 receptor. The EU authorities have responded to these new developments by issuing a new Guideline on Adjuvants in Vaccines

(EMEA/CPMP/VEG/17/03/2004) that also covers the specific toxicological aspects of this class of compounds.

Predictive value of animal studies

Predictivity is an important issue in the risk assessment of drugs: more so than in other areas such as pesticides or industrial chemicals since humans are exposed at an effect level. Questions are repeatedly raised whether toxicity found in humans had been predicted in animals. Especially in case of the withdrawal of a drug from the market because of unacceptable toxicity, the findings in the non-clinical studies before marketing authorisation will be screened again for their predictive value (or lack thereof).

The International Life Sciences Institute–Health and Environmental Science Institute (ILSI-HESI) has carried out a study with several pharmaceutical companies, analysing the toxicity described in animal studies and adverse reactions in humans, especially in cases where drugs were withdrawn from development. It appeared on the basis of 214 compounds that 71% of the human toxicity could be or could have been predicted on the basis of either rodent or non-rodent experiments. A remarkable finding is that 94% of the toxicity leading to withdrawal of the compound was found within the duration of a 30-day study. It should be taken into account, however, that there is strong selection, represented by those products that are killed in (a late stage of) development.

Carcinogenicity is another difficult aspect of predictability. Some 50% of all pharmaceuticals intended to be used long term in human patients induced tumours in rodents. The vast majority of these compound are non-genotoxic, only 6 out of 175 inducing some genotoxicity. Rodents are very sensitive to proliferation-inducers, and several mechanisms have been shown to be involved.

Example

Penfluridol, an antipsychotic drug, is pharmacologically well known to block dopamine receptors. It appeared to induce mammary tumours as well as pancreatic tumours. The mammary tumours were ascribed to the enhanced level of prolactin in the animals, though for the pancreatic tumours the involvement of prolactin was not clear at that time, and the drug was taken from the market. Subsequently, pancreatic tumours were also observed with chlorpromazine and even haloperidol, and a mechanism involving the enhanced prolactin secretion was identified. As the role of prolactin in humans is different from that in rats, the risk of a tumorigenic action of penfluridol and other antipsychotics blocking dopaminergic receptors is accepted to be low.

The induction of many tumours in rodents can be non-relevant to humans. For instance, thyroid tumours are induced by the feedback on thyroid homeostasis as a response to increased activity of the UDP-glucuronosyltransferases (UGTs) responsible for the breakdown of thyroxine. UGT activity is induced by several compounds in rodents, but not in humans. Liver tumours are often observed in long-term animal studies; these are related to an increased sensitivity in some mouse strains, and to the induction of certain liver enzymes that trigger extensive cell proliferation.

Post-marketing safety assessment

What is the role of the toxicologists after marketing authorisation? From that point, a pharmaceutical generates wide experience in an 'uncontrolled' human population (uncontrolled compared to a clinical trial). Extensive pharmacovigilance activities are more important than further toxicology studies, as the former record the experience in humans – the target population of the drug.

However, when entering the market a compound will be used in a much broader population compared to the small research populations of the clinical studies. Unexpected adverse effects may be identified that occur at a lower incidence than could be picked up in clinical studies. At that time the company has to go back to its toxicological experiments to discover the mechanism of action of such an effect. The small population of animals used in those studies might have been compensated by a higher dose and longer duration or more intense exposure to the compounds. Nevertheless, the effect may have been missed.

Information resources

Web sites where the most recent versions of guidance documents mentioned and new guidance documents can be found:

CPMP and CPMP/ICH guidance documents and Core SPC texts: http://www.emea. eu.int/index/indexh1.htm
EU guidelines (Annex V test guidelines): http://ecb.ei.jrc.it./testing-methods/
OECD guidelines: http://www1.oecd.org/ehs/test/

13

Pharmacovigilance

Sten Olsson and Ronald Meyboom

The randomised controlled clinical trial is the method of choice for the objective and quantitative demonstration of the efficacy and tolerability of a new medicine. None the less, such studies have limitations in discovering possible adverse effects that may occur, in particular those that are rare or develop after prolonged use, in combination with other drugs or perhaps due to unidentified risk factors. Clinical trials are inherently limited in duration and number of patients and, significantly, patients are selected prior to inclusion. In other words, the conditions in a trial are artificial compared with the real-life use after the introduction of a medicine. Thus, once a medicine has been approved for marketing there is a need for continued collection of information or 'pharmacovigilance'. The World Health Organization (WHO) defines pharmacovigilance as 'the science and activities relating to the detection, assessment, understanding and prevention of adverse effects and other drug-related problems'. This description illustrates that pharmacovigilance is as much about science as about monitoring and education. In particular, newly introduced medicines are given a great deal of attention in pharmacovigilance because their safety profiles are insufficiently known.

The aims of pharmacovigilance, according to the definition above, are:

- The detection as soon as possible of any hitherto unknown adverse reactions or other important drug-related problems (early warning function)
- Quantification (frequency) of the identified problem
- Benefit–harm evaluation
- Regulatory risk management (measures)
- Dissemination of information, education and prevention

Its ultimate goal is the promotion of rational and safe use of medicines.

The characteristic features of pharmacovigilance, which often make it different from formal scientific studies, are that it should cover all patients and all medicines and be continuous; it is a search for the unexpected.

Adverse reactions and other drug-related problems

With few exceptions, the therapeutic effect of a medicine is only one of many known and often unknown actions of the drug on the body. It is not surprising, therefore, that there are innumerable reasons why the administration of a medicine can lead to unexpected, unintended or even harmful results. Such drug-related problems – ranging from hyper-sensitivity reactions through interactions between drugs or with food or herbal preparations and non-compliance to acute poisoning – can be categorised in a practical way as shown in Figure 13.1. In this scheme, dose-related problems are placed above and those without an apparent dose-relationship below, whereas problems that occur during

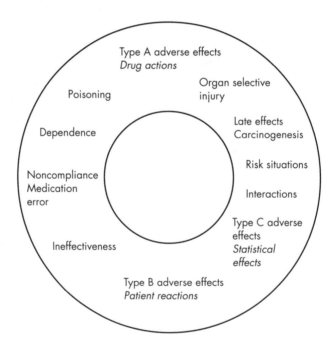

Figure 13.1 The zodiac of drug-related problems. Reproduced with permission from Meyboom RHB, Lindquist M, Egberts ACG (2000). An ABC of drug-related problems. *Drug Safety* 22(6): 415–423.

appropriate drug use are on the right side and those related to inappropriate use are on the left.

In 1969 a WHO expert group concerned with the planning of pharmacovigilance (or drug monitoring as it was called then) defined an adverse drug reaction as 'a response to a drug that is noxious and unintended and occurs at doses normally used in man'. For a medicine there is remarkably little pharmacology in this definition, but for good reasons. An indirect response, for example pseudomembranous colitis caused by a bacterial toxin during the use of an antibiotic, can be an adverse drug reaction.

Adverse reactions encompass a large collection of heterogeneous signs, symptoms, disorders and syndromes that share only the common factor that somewhere a drug has played a role in their development. Nevertheless, three main groups of adverse effects can be distinguished, Types A, B and C, based on pharmacological, pathological and epidemiological criteria.

Type A Type A adverse reactions are primarily pharmacological effects, side-effects in the true sense of the word. They are relatively easy to study because there is an exposure–response relationship (dose/time/route) and can often be reproduced experimentally in humans or animals. Type A effects are by far the most common in healthcare; they determine the tolerability of the drug and are mostly detected in clinical trials before a drug is approved for marketing. Examples are drowsiness caused by anticonvulsants, constipation caused by opioids and alopecia caused by cytotoxic drugs. Sometimes primarily pharmacological effects occur only under special circumstances; for example, congenital malformations due to maternal drug use. These can be seen as a subgroup of the A type. Other effects occur only after prolonged exposure to the drug or have a different temporal pattern.

Type B The situation for Type B, the hypersensitivity reactions is very different. These occur characteristically in only a small proportion of the users, sometimes as rare as 1 in 10 000, and may display a remarkable absence of a relationship between dose and severity. The underlying mechanism may be immunological (see Chapter 10) or may be related to an inborn error of metabolism, for example glucose-6-phosphate dehydrogenase (G6PD) deficiency, or, as is often the case, it may be of an uncertain nature. Often the reactions cannot be studied experimentally, and as a rule a rechallenge test of the patient is contraindicated.

This group includes typical drug reactions, such as anaphylaxis or agranulocytosis that otherwise have a low background frequency. Characteristic of an immunological reaction is a latency phase of about 10 days from the first exposure (sensitisation) and a rapid recurrence after (inadvertent) re-exposure. As is explained below, spontaneous reporting has been found to be particularly efficient in the monitoring of such reactions.

Type C   Type C refers to the situation in which a naturally occurring disease occurs more frequently in a drug-user population than in non-exposed patients. Although there may be a dose relationship, as a rule there is no suggestive time relationship; often there is a long interval between exposure and the occurrence of symptoms. Examples are thromboembolic disease in women using oral contraceptives or gastro-intestinal bleeding in patients using NSAIDs (nonsteroidal anti-inflammatory drugs). Type C effects are often difficult to study and remain a major challenge in pharmacoepidemiology.

Although often not considered to be a side-effect, therapeutic ineffectiveness is one of the most frequently occurring unintended responses to a medicine. Ineffectiveness is a recognised reportable event in pharmacovigilance, in particular when it occurs unexpectedly. It can have many explanations, such as inappropriate drug use (a wrong diagnosis, wrong dose or wrong duration), an interaction between drugs, or counterfeiting.

Inappropriate use is also a common explanation for other unintended effects, in particular the use of a higher dose than necessary in a sensitive patient. Medication errors are common and contribute substantially to fatal injury. Chronic inappropriate drug use, in particular in the case of dependence, can also cause serious problems. Thus, the detection of unexpected addictive properties of a new drug is of further interest in pharmacovigilance.

Epidemiology of adverse drug reactions and spontaneous reporting

There is much uncertainty about the frequency of occurrence of adverse drug reactions and other drug-related problems in society. The parameters commonly studied are the proportion of hospital admissions caused by drug-related problems and the incidence of adverse reactions in hospitalised patients. In Western countries, of patients submitted to a medical ward about 15% have a drug-related problem, i.e. adverse reaction, overdose or therapeutic failure. Serious adverse drug reactions have been demonstrated to occur in 6–7% of hospitalised patients. In roughly half of these patients the problem was considered to be preventable with better care. With a fatality rate of 0.15–0.3%, adverse drug reactions constitute a significant cause of death in society.

To date, the mainstay of pharmacovigilance has been 'spontaneous reporting' by health professionals, a system whereby case reports of adverse drug events are voluntarily submitted by health professionals and pharmaceutical companies to the national pharmacovigilance centre. In the context of spontaneous reporting a 'case report' is defined as a notification from a healthcare professional concerning a patient with a disorder suspected of being drug-induced.

Data assessment

Before reports are entered into the database at a national pharmaco-vigilance centre, they are assessed with regard to their content: the drugs administered, the event that happened, the medical history of the patient, the likelihood of a causal relationship and the possible relevance of the observation, whether it is previously known or not, its potential interest, and whether action is needed. All data that can lead to the identification of a person (patient, physician, etc.) are confidential and must not be disclosed in any circumstances. If potential reporters are not confident that confidentiality of personal details can be maintained, they are not likely to share their information and take part in the reporting programme.

Case causality assessment

A variety of questionnaires and algorithms have been proposed for assessment of the likelihood of causality. The general design of these systems is that the main question is divided into a number of sub-questions and sub-sub-questions. The basic design of criteria in case causality assessment are as shown in Table 13.1.

The answers to the sub-questions lead to scores, which are summed to yield an overall score, from which a causality category follows: for instance, 'possible', 'probable' or 'certain'. A consensus system has been developed by the WHO (see Table 13.2).

The importance attached to individual case causality assessment may be exaggerated; more important is the assessment of aggregated data, i.e. the interpretation of series of selected case reports in a given

Table 13.1 Criteria in case causality assessment

- The association in time (and place) between drug administration and event
- Pharmacological features; previous knowledge
- Medical considerations (characteristic signs and symptoms, laboratory tests, pathological findings)
- Likelihood or exclusion of other causes
- Case documentation quality

Table 13.2 WHO causality categories

Causality term	Assessment criteria
Certain	• Event or laboratory test abnormality, with plausible time relationship to drug intake • Cannot be explained by disease or other drugs • Response to withdrawal plausible (pharmacologically, pathologically) • Event definitive pharmacologically or phenomenologically (i.e. an objective and specific medical disorder or a recognised pharmacological phenomenon) • Rechallenge satisfactory, if necessary
Probable/Likely	• Event or laboratory test abnormality, with reasonable time relationship to drug intake • Unlikely to be attributed to disease or other drugs • Response to withdrawal clinically reasonable • Rechallenge not required
Possible	• Event or laboratory test abnormality, with reasonable time relationship to drug intake • Could also be explained by disease or other drugs • Information on drug withdrawal may be lacking or unclear
Unlikely	• Event or laboratory test abnormality, with a time to drug intake that makes a relationship improbable (but not impossible) • Disease or other drugs provide plausible explanations
Conditional/Unclassified	• Event or laboratory test abnormality for which more data for proper assessment are needed, or • Additional data under examination
Unassessable/Unclassifiable	• Report suggesting an adverse reaction which cannot be judged because information is insufficient or contradictory • Data cannot be supplemented or verified

context or addressing a given question, in particular signal detection. The combination of many 'possible' case reports may produce a 'probable' signal.

From data to information

The next step in data assessment concerns data interpretation, the transformation of the input data into meaningful information:

- Signal detection
- Frequency estimation
- Structured study of series of selected case reports (clinicopathological)
- Quantitative studies and data mining in large databases

Signal detection

Pharmacovigilance is looking for the unexpected. In terms of spontaneous reporting, WHO has defined a 'signal' as 'reported information on a possible causal relationship between a drug and an adverse event, the relationship being unknown or insufficiently documented previously'. The principle underlying signal detection in spontaneous reporting is simply that *when different reporters independently report the same unknown and unexpected adverse experiences with a drug, this may constitute valuable information.* The reporting system is particularly efficient when case reports concern a single suspected drug and a characteristic objective clinical event that is unlikely to be related to the medical history of the patient (e.g. the indication for use), with a low background frequency and occurring in a suggestive time relationship. In addition, the quality and detail of the documentation of the observations in the case reports have to be taken into account. In practice, however, the situation is often different.

There are two approaches to the identification of suspected adverse drug reactions of possible interest. On the one hand, every time a new report is received and entered into the adverse reaction database, the possible importance of the observation is considered (e.g. established or unknown reaction, number of previous similar reports). On the other hand, in particular at centres with large databases and in international pharmacovigilance, automated quantitative systems are used to measure the unexpectedness of reported drug–event combinations against the background of the database.

Addressing the evidence in and importance of a signal

The major function of spontaneous reporting is the provision of early warnings about possible drug-related problems. A challenging responsibility of professionals working in pharmacovigilance is that of identifying associations of potential importance as early as possible, while at the same time avoiding false alarms.

A characteristic function of a national pharmacovigilance centre is that of making an expert assessment of the available data regarding effectiveness and risk of medicines and taking a position, even though the evidence is still uncertain. This decision requires great care and professional expertise (Table 13.3). Obviously the number of victims of a serious new adverse reaction must be kept to the minimum, but a false alarm can harm a valuable product, the company, the national centre, and patients under treatment.

Table 13.3 Criteria for assessing the evidence in a signal in pharmacovigilance

Criterion	Explanation
Quantitative	
Strength of the association	The number of case reports; statistical disproportionality; low background frequency
Qualitative	
Consistency of the data	The general presence of a characteristic feature or pattern and the absence or rarity of converse findings
Exposure–response relationship	Site, timing, dosage-response, reversibility
Biological plausibility of the hypothesis	Pharmacological and pathological mechanisms
Experimental findings	A positive rechallenge; drug-dependent antibodies; high blood or tissue drug concentration; abnormal metabolites; diagnostic markers
Analogy	Previous experience with related or other drug; event known to be frequently drug-induced
Nature and quality of the data	Characteristic nature and objectivity of the event; accuracy and validity of documentation; case causality assessment

Users of data

Pharmacovigilance is a matter of producing information intended to improve the rational and safe use of medicines. A variety of users of the results of spontaneous reporting can be identified – health authorities, pharmaceutical companies, academia, drug information centres, national drug bulletins and, of course, healthcare professionals and patients. In the past the use of spontaneous reporting databases was restricted to regulators and companies, but nowadays in many countries the data are easily or even freely accessible.

Strengths and weaknesses of spontaneous reporting

Worldwide experience has shown that spontaneous reporting is reasonably efficient in the signalling of new adverse effects. Its advantages are that it is continuous and rapid, covers many patients, many drugs and many adverse events at the same time, works in countries with a less structured healthcare system, and is relatively cheap. It normally covers the whole population for an unlimited period, and may therefore have a better chance of revealing rare, serious adverse drug recations.

Limitations of the system follow from the uncertain nature of the data and the lack of descriptive elements in the case reports, making an assessment of causality difficult. It suffers from under-reporting and may fail in providing proof, measuring the frequency of an adverse effect and comparing the safety of different medicines.

There is vast under-reporting of adverse reaction reports worldwide: that is, not all adverse reactions that occur are reported to pharmacovigilance centres. However, since the extent of under-reporting is unknown and variable, it is difficult to adjust for. It has been estimated that, even in countries with a long tradition of adverse reaction reporting, often no more than about 10% of adverse drug reactions are notified to the pharmacovigilance centres. Under-reporting is influenced by a number of factors: reporting is higher for new drugs than for old; serious reactions are specifically requested to be reported and their under-reporting is less. Similarly, type B reactions are reported more commonly than their share of events in practice.

Publicity also plays a major role in reporting: reporting is affected by promotional claims of the drug sponsor, and publicity for a specific drug-related problem triggers further reports. Reporting is affected by general publicity around the adverse reaction reporting scheme.

To stimulate the collection of adverse drug reaction reports, many countries have introduced legislation making it mandatory for pharmaceutical companies to report all cases that have become known to them to the regulatory authority. Serious, unexpected reactions have to be reported within 15 days. The severity of a problem is determined by both the seriousness and the frequency of the adverse reaction. Unfortunately, the possibilities of spontaneous reporting for measuring the quantitative aspects of a problem and for comparing the safety of different drugs are limited. In addition to under-reporting, which is difficult to adjust for, drug sales or consumption data – the denominator – may be missing or difficult to interpret. Since the balance of the benefit and harm of a drug often depends strongly upon the incidence of adverse reactions, this weakness of spontaneous reporting is a major concern to risk management and regulatory decision making. The pros and cons are listed in Table 13.4.

Depite the obvious limitations, spontaneous reporting is still the primary method used at pharmacovigilance centres around the world as a first-line safeguard. However, a further study is often needed to prove the connection, identify risk factors and measure the frequency of the reaction. This is the more important since, in addition to seriousness, the importance and acceptability of an adverse reaction is determined by its frequency.

The major function of spontaneous reporting is the provision of early warnings about possible drug-related problems. Often such hypotheses need to be tested and verified, using experimental or epidemiological methods.

In drug regulation, a reaction is considered to be serious when it

Table 13.4 Pros and cons of spontaneous reporting

Advantages	Limitations
• Effective	• Mainly reports of suspicions
• 'All' patients; 'all' drugs; many different adverse reactions	• Low participation of practitioners
• Continuous	• Bias
• Rapid	• Insensitive to type C adverse effects
• Cheap	• Drug consumption data rarely available
• Works also in healthcare system with little structure	• No quantitative assessment
• No specialist training for practitioners needed	• Comparison between drugs is difficult
	• Often further study needed (proof, frequency, risk factors, mechanism)

is life-threatening or fatal, leads to hospitalisation or prolongation of hospitalisation, causes persistent incapacity or disability, or includes congenital malformations.

Other methods in pharmacovigilance

In addition to spontaneous reporting, several other methods have been developed for use in pharmacovigilance, in particular prescription event monitoring (PEM) and case–control studies. So far, however, these methods are being used not in routine drug safety monitoring but in special situations. PEM is employed to monitor the safety of selected new drugs, for example those with a novel mechanism of action. Through a central registry of prescriptions (or collaboration with pharmacists), doctors prescribing the selected medicines are identified. They are provided with a special reporting form on which they are asked to record any medical occurrences happening to their patient while using the drug of interest. All events, whether suspected to be related to drug treatment or not, are recorded. The recorded events are computerised and the profile of events for all individuals being monitored (the cohort) is compared with a control. The control population may be a cohort of patients exposed to a similar drug. Normally cohorts of approximately 10 000 patients are collected for this kind of study. PEM may identify unexpected associations that were not identified in spontaneous reporting systems and also provide incidence rates, since both the number of observed events and the number of exposed patients are known. The size of the study normally does not allow identification of very rare reactions or those occurring after prolonged use because of the restricted time window.

In case–control studies, individuals affected by the adverse event being studied are identified (cases). Each case is matched with several disease-free control patients randomly recruited from the study base (controls). Both cases and controls are investigated regarding their exposure to possible causative agents prior to the occurrence of the event. The odds ratio, which is an estimate of the relative risk, is calculated on the basis of exposure data. If a given exposure is demonstrated to be more common among cases than among controls, the relative risk is >1. The case–control design allows many different exposures to be investigated simultaneously. It is commonly used to study rare outcomes and is often employed to confirm hypotheses.

International pharmacovigilance and the WHO Collaborating Centre for International Pharmacovigilance in Uppsala

In 1968 the World Health Organization introduced the WHO Programme for International Drug Monitoring. The basic idea was that the pooling and integral assessment of national data from as many countries as possible would greatly accelerate the detection of new drug hazards. Originally starting with 10 national centres, the numbers of participating centres and of case reports increased steadily, with today 79 participating countries and over 250 000 case reports per year. At regular intervals national centres send summaries of case reports in a standardised electronic format.

The operational aspects of this programme are now maintained by the WHO Collaborating Centre for International Drug Monitoring in Uppsala, Sweden, a non-profit foundation. The WHO Uppsala Monitoring Centre (UMC), as it is often called, has an international administrative board; the WHO has retained responsibility for policy issues.

The global database of the UMC is used in many different ways. The large and unique data collection of the UMC is an easily and frequently used first-line reference source, not only for signal detection but also for early hypothesis strengthening. The finding that the same unknown suspected adverse reaction to a new drug is observed in several countries around the world can be a strong argument supporting a causal relationship.

Analysis of a series of reports of patients experiencing the same adverse reaction after having been exposed to the same drug often allows conclusions to be drawn about risk factors contributing to the development of the reaction, which is helpful when giving advice on how to avoid the problem in the future.

Taking regulatory measures is the responsibility of the competent authorities in the participating countries.

Other forms of international collaboration: EudraVigilance

In the context of the further development of centralised regulation in the countries of the European Union, the EudraVigilance network (http://www.eudravigilance.org/highres.htm) has been developed by the European Agency for the Evaluation of Medicinal Products (EMEA) National centres and pharmaceutical companies in member states are required to report serious suspected adverse reactions to 'centrally licensed' medicines. The scientific advisory committee to the European Commission, the Committee for Medicinal Products for Human Use (CHMP), has a pharmacovigilance working party with representatives from all member countries. The country that served as a rapporteur, making the assessment of documentation prior to the granting of a product licence, is also responsible for the analysis of post-marketing data. Other international bodies, such as CIOMS (the Council for International Organisations of Medical Sciences; www.cioms.ch) and ICH (the International Conference on Harmonisation of Technical Requirements for Registration of Pharmaceuticals; www.ich.org), are major players in the further collaboration and harmonisation in pharmacovigilance around the world.

Communicating drug safety information

There is no point in learning more about drug-related risks unless the knowledge is brought to the attention of the professionals who are in a position to make decisions about drug selection and drug therapy. It has been shown that approximately 50% of adverse reactions leading to hospitalisation can be considered avoidable, which means that if the existing knowledge had been appreciated and acted upon properly by the prescriber and the patient, the adverse drug reaction would not have occurred. There is often a wide gap between what is known about drug risks by pharmacovigilance experts and by healthcare practitioners and patients. One of the greatest challenges in pharmacovigilance is to bring information about effectiveness and risks of medicines to all the stakeholders in society who need to know and act on the information in a rational way.

Further reading

Edwards IR, Biriell C (1994). Harmonisation in pharmacovigilance. *Drug Safety* 10(2): 93–102.
Meyboom RH, Lindquist M, Flygare AK, Biriell C, Edwards IR (2000). The value of reporting therapeutic ineffectiveness as an adverse drug reaction. *Drug Safety* 23(2): 95–99.

Information resources

WHO Collaborating Centre for International Drug Monitoring: www.who-umc.org
The International Society of Pharmacovigilance, ISoP: www.isoponline.org
The International Society of Pharmacoepidemiology, ISPE: www.pharmacoepi.org

Index

Page numbers in *italic* refer to figures and tables.